Canyon Country Geology

for the
Layman and Rockhound

A guide to understanding the
unique and spectacular geology
of the canyon country of southeastern
Utah and vicinity, with a special
section on rockhounding

By
F.A. Barnes

ANOTHER
CANYON COUNTRY
GUIDE BOOK

1978
Wasatch Publishers, Inc.

This book is the ELEVENTH in a series of practical guides to travel and recreation in the scenic Colorado Plateau region of the Four Corners states

All written material, maps, charts and photographs are by F.A. Barnes unless otherwise credited.

Seventh Printing 1993

Artwork
by
Susann Bléu

© 1978
Wasatch Publishers
4460 Ashford Drive
Salt Lake City, Utah, 84124
ISBN 0-915272-17-2
Library of Congress No. 77-95050

CONTENTS

Fran Barnes
Photo by Terby Barnes

FOREWORD

Books on canyon country geology should rightfully be written by professional geologists who are familiar with this unique region. In fact, knowledgeable geologic scientists have published great quantities of material on the region, but most of it in a form useful only to other geologists.

A few booklets have been issued on local aspects of canyon country geology, such as its national parks or along its river gorges, but there is nothing in print of a regional scope, written for the many people who are fascinated by the region's geologic diversity and beauty and who want to know more about what they are seeing.

Thus, the task of writing a non-technical canyon country geology book fell to me by default. My qualifications for attempting this task do not include a degree in geology, but do include more than two decades of scientific research, applied engineering and considerable experience at literature research, technical writing and editing. I also have a broad background of study in various sciences, and a keen personal interest in the human and natural history of the general Four Corners region, an interest I have pursued assiduously for the past ten years.

In compiling the information in this book I used many sources, including geology textbooks, technical books, maps and scientific reports on canyon country geology, discussions with eminent and knowledgeable geologists and—last but not least—my own trained observational capacities.

My literature research was at times quite frustrating. Geology, as with almost all sciences, is always in a state of change, as new concepts are proposed, examined and tested, then accepted or rejected. Today's widely accepted "theories" and "laws" are often the scientific heresy of yesteryear. Further, not all scientists accept new concepts at the same time, and outmoded ideas get into print, and stay in print, for a long time.

Thus, at any point in time, the existing literature on any particular scientific subject represents a confusing mixture of old and new, of tried-and-true and wild-eyed hypothesis. The field of geology is no exception. For the last ten years all aspects of geology, plus many life sciences, have been in a state of rapid flux due to the firm verification of one such wild-eyed hypothesis, that of continental drift and sea-floor spreading.

This concept, alone, has created turmoil and dramatic changes in geologic theory, but has also been responsible for making a cohesive whole out of great quantities of poorly understood and seemingly contradictory research data. This process of unification is still continuing at a rapid pace in all of the geologic and biologic sciences.

While writing this book, I was forced to make compromises. The full, detailed story of canyon country geology is far too complicated for a book even ten times the size of this one. Further, such a book would be of interest to very few people, primarily a few geologists, and it would not serve the purpose for which this book is intended.

I therefore made compromises. I simplified some geologic processes for the sake of understanding by readers untrained in the sciences. I sometimes had to choose between conflicting statistics, and I eliminated mention of several minor factors of the type that are vital to scientists, but meaningless to laymen. I ignored whole fields of specialization and complex terminology in favor of trying to promote a broad, overall understanding among non-professional readers.

The results of my compromises may horrify some professional geologists, especially those who may not be accustomed to translating their subjects into terms understandable by non-professionals. But I assure the readers of this book that its contents are quite accurate enough to serve the book's intended purpose, that is, to be instructive and useful for the average person — whether canyon country visitor or resident — and to answer in an understandable manner the kinds of questions that are brought to mind by the awesomely beautiful, diverse and unique geologic features of canyon country.

The purpose of this book is to answer the kind of general-interest questions that are almost universally ignored in the serious writings of geologic scientists or, if they are answered, the answers are couched in scientific terms that are incomprehensible to the non-professional.

My apologies to professional geologists who may have some interest in canyon country. I have great respect for your efforts, and invite your constructive comments on mine, but my assumed task was to interpret your complex findings and theories, and to translate these complexities into terms understandable by many people, the people who want to know more about canyon country geology and the myriad unique geologic features found here, without taking the time to acquire a degree in geology. I hope I have succeeded.

6

Visitor Center, Capitol Reef National Park.

CANYON COUNTRY

UTAH

COLORADO

COLORADO

PLATEAU

NEW

ARIZONA

MEXICO

FIGURE 3

INTRODUCTION

REGION COVERED

In general, this book describes the spectacular and unique geology of the canyon country region of southeastern Utah. Other books in the "Canyon Country" series cover other aspects of travel, outdoor recreation and the natural history of this same region, with each book defining the extent of "Canyon Country" for its own purpose and scope. These perimeters are either obvious natural features, such as river gorges, or artificial boundaries, such as highways or state lines.

From a geologic viewpoint, canyon country was not so easy to define or limit. Common usage of the term "Canyon Country" provided little help. The residents of several southeastern Utah communities commonly apply the term to their own areas, and the term could also rightfully be applied to parts of Arizona and Colorado.

Similarly, canyon country as a geologic region could not readily be defined by the use of obvious natural features and artificial boundaries. Its geologic roots are too deep for that. Another approach had to be used.

After considerable study, the region shaded in Figure 1, on the inside-front cover, was chosen for description. State or federal highways were used for much of the regional perimeter for two reasons: they are accessible and obvious to travelers; and these particular highways just happen to closely parallel major changes in geologic formations. The straight-line boundaries were also placed as shown for two reasons: for traveler convenience and orientation; and to include surface features and areas that are a vital part of the canyon country past and present.

Thus, for the purposes of this book, the canyon country perimeter is defined as follows: a straight line between Grand Junction, Colorado, and Price, Utah; Utah 10 between Price and Interstate 70; a straight line between the U 10/I 70 junction and Glen Canyon City, Utah; U.S. 89 between Glen Canyon City and the U.S. 89/U.S. 160 junction in Arizona; U.S. 160 between the U.S. 89/U.S. 160 junction and Cortez, Colorado; a straight line between Cortez and Grand Junction.

Of course, most of the major geologic formations found within this outlined region can also be found outside of the region.

The painted-desert Chinle Formation dominates Petrified Forest National Park in Arizona, and is also exposed many other places in the Four Corners states. The soaring cliffs of Zion National Park are Navajo Sandstone, and this same spectacular rock also appears in northwestern Colorado, southwestern Wyoming and central Utah, although in some localities under a different name.

The rocks and sediments of the Morrison Formation occur in states to the east and west of the Rocky Mountains, and from northern Arizona and New Mexico to the Canadian border, with surface exposures in many places.

Other rock layers that appear in canyon country, such as the Dakota Formation, occur over much of the North American continent, but in most cases are deep beneath the surface, or are exposed in few places. Only in canyon country and in the Grand Canyon is such a wide variety of ancient geologic formations exposed to easy viewing and appreciation.

The exposed geologic layers in these two unique and spectacular regions supplement each other. Exposed canyon country formations begin about where Grand Canyon formations end, with little overlap. The general relationships between the geologic formations exposed in various Colorado Plateau national parks, monuments and recreation areas are shown in Figure 2 on the inside-back cover.

The region thus defined as canyon country has three major attributes, all important to the purposes of this book:
1. The region is all within the geologic province called the Colorado Plateau, and hence has acted, and been affected, as a geologic unit for a long period of time, in particular the last 500 million years. See Figure 3.
2. The same general geologic strata occur throughout the region, with few exceptions, although exposures of these strata vary from area to area.
3. These geologic formations, where they surface, are generally not hidden by covering soils, forests and other vegetation, but are exposed to ready viewing.

These three attributes have combined to produce the startling, diverse and unusual geology of canyon country, and to make that novelty and beauty accessible and visible to those who visit or dwell within the region. They have created canyon country as it is today, and left it an open, natural geology textbook, waiting to be read and appreciated.

The purpose of this Canyon Country guidebook is to help those untrained in the science of geology to read this natural textbook and thus to enhance their understanding and appreciation of the many wonders to be seen in the canyon country of southeastern Utah and vicinity.

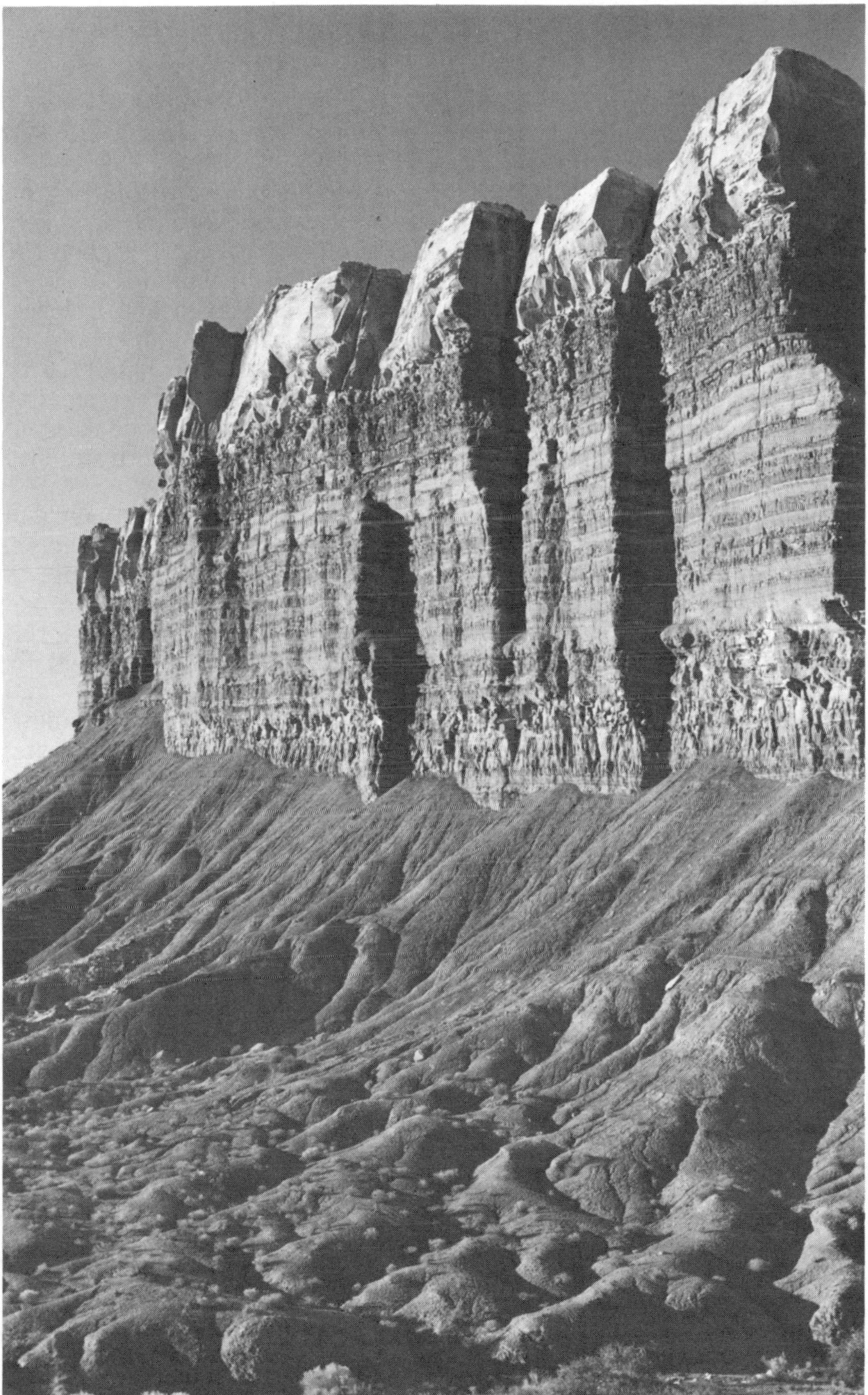

Capitol Reef National Park.

REGIONAL GEOGRAPHY

Canyon country, as defined, is essentially an elevated desert plateau with extremes in elevation provided by mountain peaks and deeply cut river gorges.

In addition to the three distinct mountain ranges within canyon country, the Abajos, La Sals and Henrys, several major uplifts have created other high areas within this immense plateau. These are the Circle Cliffs, Monument, Uncompahgre and San Rafael uplifts.

Four major rivers, the Colorado, Green, San Juan and Dolores, have produced an intricate system of gorges and tributary canyons that cut deeply into the desert plateau. The Escalante, Fremont, Dirty Devil, Muddy, San Rafael, Price, Paria, and San Miguel are smaller tributary rivers that also contribute significant canyons, as do many of the countless perennial and intermittent streams that drain into canyon country rivers.

Altogether, these rivers and streams have produced a vast and complex canyon system that drains the entire canyon country region, reveals its underlying geologic layers and gives the region its name.

Between its highs and lows, other major geographic features add diversity to the high desert plateau of canyon country. Broad, colorful deserts contain huge outcroppings of sandstone and are walled by looming escarpments of multi-hued rock. Vivid painted deserts and eroding hills of gray shale are accented by higher terraces of yellow-brown sandstone and by sinuous lines of bright green trees bordering meandering watercourses.

Long, soaring clifflines and the highlands beyond them are slashed by canyon systems as complex as the Grand Canyon, although smaller in size. Elevated, slender mesas, peninsulas and spires jut out from many such clifflines, still further subdividing the deserts and canyon systems below the cliffs.

Unique salt valleys slash into the La Sal mountain foothills. Broad, open plains of ancient ocean sediments provide another type of desert, and in some places igneous rocks and glacial moraines break the pattern of water- and wind-deposited sediments.

Many small lakes, few of them natural, dot canyon country highlands and mountains, and one immense reservoir, Lake Powell, lies in the heart of a vast region of barren slickrock.

In addition to the reservoirs, other manmade features contribute to canyon country geography. Several towns and small communities are connected by a network of highways. Roads and trails penetrate the backcountry for various purposes. In some areas, natural terrain has been converted   to cultivated cropland, while other areas show the massive scarring typical of prospecting and mining activities. Large ranches are apparent in some of the mountain foothills, railroads cut across the region in two places and the tall steel towers of high-tension power lines cross canyon country in several places.

Altogether, the geography of canyon country is quite diverse, and to date its countless fascinating and beautiful surface features are relatively unmarred by human activities. The following chapters discuss many of these surface features and the geologic events and processes that formed them.

GENERAL GEOLOGY

The geology of canyon country is very complex. Its surface features were produced by more than 500 million years of geologic activity, some of it localized, some regional, some continental and some global. Of necessity, this book must be limited to a general discussion of the major geologic events and processes involved. The literature listed in the Further Reading chapter in the back of this book provides more details for those who wish to delve more deeply into the mystery and intricacy of canyon country geology than the scope of this book permits.

Until the early 1960s geologists had to be content simply to describe most major geologic events, including those in canyon country, without having any idea about the fundamental causes of these events. They could report in great detail the history of mountain ranges, uplifts and volcanoes. They could tell of invading seas and record patterns of faulting and earthquakes. They could describe how vast regions were buckled into accordion-pleated basin-and-range provinces, and other even larger continental areas were covered by invading glaciers or vast tropical swamps.

13

Still other scientists could tell about worldwide changes in climate, ocean temperature, sea-level and the development and extinction of countless species of life. They could report strange differences and similarities between life forms and geologic forms on different continents, and note the earlier existence of tropical and subtropical animals in what are now arctic regions. They could tell about vast, arid deserts where teeming swamps had once been.

Geologists and other scientists who have studied canyon country have carefully described all of these events and many, many more, and have described in great detail how they relate to each other. But until the last decade or so they have been unable to say why these events have taken place. They were able to report WHAT events had shaped the geologic wonderland of canyon country, but not WHY these events had occurred. Their image of this land was thus incomplete, only half the story.

Within the last decade or so, however, geology and many other fields of earth science have entered a new phase in which many of their fundamental questions about basic causes are being answered. The newly developed, and still developing, concept of continental drift has provided a wealth of new knowledge which has led to an understanding of why certain geologic, biologic and other events happen.

Geologists who study canyon country are now beginning to relate the global events of continental drift to the regional events of western North America and to the Colorado Plateau province. A general understanding of continental drift concepts is essential to an understanding of canyon country geology, because the basic forces that have shaped this complex region for the last half-billion years are continental drift events, plus the secondary effects these events have in turn produced.

Following chapters of this book describe general continental drift theory and some of the major continental drift events that have affected canyon country geology. More about this fascinating and highly important concept can be found in several of the references listed under Further Reading.

Despite more than one hundred years of sporadic research into the geology of canyon country, plus current efforts to relate existing knowledge to the developing concept of continental drift, there is still much to be learned about this fascinating land and the forces that have shaped it. New research methods are being developed and applied. New information is being produced. New relationships between geologic events are being hypothesized.

Unfortunately, much of the current research is being oriented toward the search for minerals. An understanding of the region's geologic history can lead directly to the finding of deposits of fossil fuels, uranium and other minerals. Other research, however, is oriented toward the acquisition of general knowledge, or toward purposes that are less destructive to the unique geologic features of canyon country.

One of the first things noticed in canyon country is that its exposed rocks and sediments are stratified, or separated into distinct layers, with each layer having its own special characteristics.

The next thing noticed is that in many places these geologic strata are distorted, or deformed from their original near-horizontal position. Some such deformations are extreme, and obviously involved enormous forces.

Then, depending upon how widely canyon country is explored, many other major and minor geologic features are noticed. Some of these are quite obvious and common, some are not so apparent and are rare, but all are worthy of note and contribute to the unique form and beauty of canyon country.

The following chapters of this book approach canyon country geology in this manner. After a chapter describing the geologic time scale and the general flow of events on that scale over the past half-billion years, other chapters describe the various geologic strata in canyon country, the events that deformed them and the major and minor features that resulted. A final chapter tells how to apply this general knowledge of canyon country geology to the collection of mineral specimens.

It is not the purpose of this book to provide instructions in basic geology and mineralogy. That kind of information can be found in many other publications. A few of these are listed under Further Reading. Certain basic geologic terms for which there are no common words are defined and used, however, in this book.

Succeeding chapters make reference to various features, places, travel routes and communities. The names for these are as they appear on official state highway maps, U.S. Geological Survey maps, other Canyon Country guidebooks and maps, and various literature references. For the titles of other useful Canyon Country guidebooks and maps see the list near the back of this book.

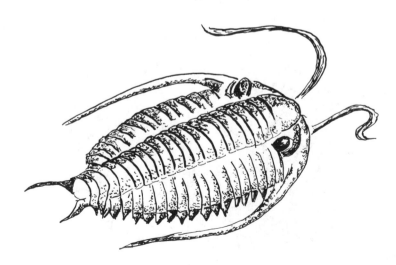

GEOLOGIC HISTORY

GEOLOGIC TIME SCALE

A generalized time scale for our planet's history has been standardized and adopted by all earth sciences, including geology. This scale is shown in Figure 4, together with brief explanations of the reasons for the various subdivisions of time. The long Precambrian era also has subdivisions, but names for these have not yet been standardized. Figure 4 is on the back cover.

Following are brief summaries of the geologic history within canyon country during the various periods shown in Figure 4. More details about some of the major events noted in these summaries appear in a later chapter entitled "Major Shaping Events."

The ages given for each geologic period are only approximate, but are accurate enough for the purpose of describing the very slow geologic and biologic processes that define and occur within these periods. Each of the periods shown also has other subdivisions of time. Most of these are useful only to scientists, but for the purposes of this book, the two most recent periods — Tertiary and Quaternary — are described in terms of "epochs," or their next smallest time subdivisions.

In the next chapter, another subdivision of the geologic time scale called "rock units," or rock strata, is discussed.

GEOLOGIC PERIODS

Precambrian Era (570 million years ago and earlier). This term is applied to the entire age of the Earth from about 570 MY ago to its creation about 4,635 MY ago. Ancient Precambrian rock is the foundation under more recent canyon country rock strata, although in a few places Precambrian rock can be found on the surface. During the late Precambrian, sediments deposited by shallow seas later changed into the harder, more durable granite-like form that is typical of Precambrian rock. Certain geologic events that took place during this immensely long era, such as faulting, probably affected more recent events to some extent. For example, the same underlying fault system in Precambrian rock may have contributed to the formation of both the Ancestral Rocky Mountains and the present Rocky Mountains.

Cambrian (500 to 570 MY ago). During this period, a shallow sea invaded Utah from the west. The canyon country region was somewhat higher, although not very much, and was inundated last. The seas were warm, because Utah was equatorial at that time. Erosion later removed some of the Cambrian deposits from canyon country, but rock of this period still remains. Cambrian rock is not exposed on the surface in canyon country, but is found below the surface almost everywhere.

Ordovician (435 to 500 MY). By this period, the Cambrian sea had retreated from canyon country, although it remained in western Utah. Because canyon country was exposed to erosion during and subsequent to the Ordovician period, no new rock layers were deposited, or if some were, they were completely removed by later erosion. Thus, the underground strata in canyon country contain no Ordovician rock. Because the equator ran through Utah during this period, the seas in western Utah were warm, and rich with primitive sea life. Neither plant nor animal life had developed on land at this time.

Silurian (395 to 435 MY). It is not known positively whether the Ordovician sea that covered western Utah re-invaded canyon country during the Silurian period, or whether the gentle erosion of the late Ordovician period continued. If the sea did return, all deposits laid down then were later removed by erosion, because canyon country strata do not contain Silurian period rock.

Devonian (345 to 395 MY). The erosion begun during the Silurian period continued through half of the Devonian within canyon country, leaving no deposits representing this time. The gap in canyon country rock strata between the Cambrian and mid-Devonian period is called

17

an "unconformity." Erosion during later periods left other unconformities in canyon country rock strata. During the last half of the Devonian, shallow seas once again covered canyon country most of the time, but there were also intervals of limited erosion. As a result, rock strata representing the upper Devonian period do exist in canyon country, but only beneath the surface.

Mississippian (325 to 345 MY). The late Devonian sea that invaded canyon country deepened still further during the Mississippian period. The marine life that thrived in the warm equatorial sea left deep deposits of limestone to represent the first half of this period. Later deposits were removed by late-Mississippian, early Pennsylvanian erosion. Mississippian period rock is found under most of canyon country, but is exposed on the surface in very few places.

Pennsylvanian (280 to 325 MY). The western sea began another slow invasion of canyon country early in the Pennsylvanian period. Other major geologic events that helped shape the region also occurred during this period. Early in the period, an uplift on the western border of canyon country resisted the sea. Geologists named this the Emery Uplift, or Paiute Platform.

In the middle of the period, as the sea slowly deepened, an immense uplift to the east began forming the Ancestral Rockies. Part of that extensive range of mountains angled across the Utah-Colorado border. Geologists named this the Uncompahgre Uplift. As the uplift rose, it divided a great shallow basin of the sea. The southwestern half of this great oval depression, called Paradox Basin, ultimately extended from northwestern New Mexico, to Price, Utah. As the sea level gradually rose, this basin sank lower and accumulated thousands of feet of salts, evaporated from the sea water, plus sediments from nearby higher ground.

During the last third of the Pennsylvanian period, the encroaching sea finally completely inundated Paradox Basin and all of canyon country, including the initial Uncompahgre Uplift. Later in the period, the Uncompahgre lifted still again, and erosion during this and subsequent periods removed all sedimentary deposits from its highlands, exposing its Precambrian rock spine. Erosion also removed all of the earlier Pennsylvanian deposits from the Emery Uplift, but most of the canyon country region received thick sedimentary deposits from these and other highlands during the Pennsylvanian period. As this period ended, the sea withdrew from canyon country. An unconformity about this time indicates an interval of erosion of Pennsylvanian marine deposits before the beginning of the freshwater Permian sediments that covered much of canyon country.

Upper Pennsylvanian rock strata are exposed many places in canyon country.

It is interesting to note that from the late Precambrian, and perhaps earlier, to the end of the Pennsylvanian period, canyon country was essentially a coastal tropical or subtropical region, sometimes under water, sometimes above, as the sea level rose and fell.

Some of the reasons for this fluctuating sea level are discussed in a later chapter.

Permian (225 to 280 MY). By the Permian period, the sea had retreated to the west, but not very far and not permanently. One bay extended eastward almost to the present town of Moab, leaving extensive marine deposits. To the east of canyon country, sediments from the immense mountainous highlands of the Uncompahgre Uplift spread toward the west and southwest, ultimately reaching into northwestern Arizona. At various times during the Permian, rock and sediments from the Uncompahgre interfingered with marine deposits, as the sea advanced into canyon country, then retreated.

During the mid-Permian, the sea twice intruded into canyon country from the northwest, each time leaving thick deposits of seashore sand. Heavy sediments from the Uncompahgre deposited immediately to the southwest of this uplift depressed and disturbed the underlying Pennsylvanian salt beds of the Paradox Basin, causing the subsurface salt flows that ultimately led to the formation of canyon country's several "salt valleys."

During the Permian, canyon country climate was hot and arid, with the equator now angling across Texas. Winds carried great masses of sand from the Uncompahgre and adjacent lowlands, and formed a vast area of shifting dunes in the southern part of canyon country. Rock formed from these dunes makes up the sheer cliffs of Monument Valley and Canyon de Chelly.

By the end of the Permian period, the Emery Uplift in western canyon country was buried under Permian sediments, and the mighty Uncompahgre range was eroded down to an expanse of low, rolling hills, after covering most of the canyon country region with thick beds of red sediments. The Ancestral Rockies were essentially gone, worn down by many millions of years of wind and water erosion. The region to the west and southwest of the Uncompahgre, on to the sea in western Utah, was largely low open plains, with few distinguishing geologic features. The Uncompahgre lifted for a third time as the Permian ended. Very late Permian and early Triassic rock strata are missing in canyon country, due to an interval of erosion.

Permian rock strata are exposed in many areas of canyon country, and comprise a very important part of its scenic beauty.

Triassic (190 to 225 MY). Because of the late-Permian, early-Triassic unconformity, the earliest Triassic rocks are missing from canyon country rock strata. The oldest remaining sediments of this period were deposited by sluggish fresh water streams. The Permian sea still covered much of western Utah and formed vast tidal mudflats in western canyon country. As the sea level rose and fell, these low tidelands shifted east, then west, and stream-borne sediments from higher land to the east interlayered with shallow marine deposits from the west. The climate was warm and humid, and between the western sea and the low remnants of the Uncompahgre Uplift and other ranges farther south and east, vast floodplains and mudflats, laced with streams and brackish ponds, provided an ideal setting for lush vegetation and the reptiles and amphibians that lived there. Eventually, the sea retreated out of canyon country, and freshwater sediments dominated.

The mid-Triassic period is also missing in canyon country rock strata, and again represents an interval of erosion. Then, deposits of somewhat different nature began to cover the whole region. The climate was temperate and the basic deposits were on a vast river flood-plain, with sediments from highlands to the east, but with a liberal mixture of volcanic ash from southern Arizona added. This interval of sedimentation completely covered the remnants of the Uncompahgre Uplift, with mid-Triassic material deposited directly onto the exposed Precambrian rock spine of the uplift. These deposits were somewhat deformed by a fourth uplift of the Uncompahgre during this interval.

During the last part of the Triassic period, climatic changes turned canyon country into a vast arid desert of shifting sand dunes. For one interval during this time, the region was covered by streams, shallow lakes and mudflats. Then the desert dunes returned, to dominate the whole region on into the early Jurassic period.

Triassic period rock strata are widely exposed in canyon country and provide some of its most spectacular scenery.

Jurassic period (136 to 190 MY). The late Triassic-early Jurassic desert dunes were extensively eroded in the canyon country region during the first part of the Jurassic period, leaving an unconformity in the rock record. Then a shallow sea invaded the region, reaching down from Canada to the southwestern corner of Utah and covering the western part of canyon country. As this sea gradually retreated to the north, it left tideland and mudflat deposits. These and similar coastal deposits covered the whole region. As the sea left canyon country, much of it was once again covered by sand dunes. The desert phased into low sea-coast tidal lagoons and mudflats in the western part of canyon country.

About this time, two major geologic events began which affected the canyon country region in many ways throughout succeeding periods. The North American continent began to separate from Europe, and rock strata in western Utah and adjoining states began to thrust eastward, overriding other strata and creating a mountainous highland. These two events were probably related.

Next, a short interval of erosion was followed by a second intrusion of the sea from the north. As the sea again retreated, it left seashore and tidal flat deposits. As the sea lost ground, desert dunes advanced from the south and east. The entire region, plus parts of several adjoining states, was next covered by sediments from streams, mudflats and shallow lakes. During this interval, considerable volcanic ash fell on the region, coming from sources to the northwest. The water-borne sediments came principally from highlands to the south-

west, in Arizona, with some from the recently formed highlands to the west.

These last Jurassic deposits were made during a time when the region's climate and wet surface conditions were ideal for the giant reptiles that roamed this continent for most of the Mesozoic era. Their fossilized bones are common in this last Jurassic period rock formation.

Jurassic strata are exposed throughout canyon country and include some of the region's most colorful rock formations.

It is interesting to note that from Precambrian time to the end of the Jurassic period, canyon country was either under shallow seas, was low coastal flats or was low, open land with few outstanding features. The only major exception was the Uncompahgre uplift which raised highlands in the eastern part of canyon country during the Pennsylvanian, Permian and Triassic periods, highlands that were eroded down, then covered over during the late Triassic. While the foundations for present canyon country features were being laid during this half-billion years or more, its beauty, color and variety of geologic form were still buried deep beneath the surface of flat, open country that was very little above sea level, or beneath invading seas.

Cretaceous (65 to 136 MY). The meandering streams, river flood plains and shallow lakes of the late Triassic period continued into the early Cretaceous, although distinct differences in the colors and textures of these adjacent deposits indicate climatic and other changes. A brief interval of erosion was followed by the gradual invasion of the region by a sea. As this inland sea advanced from the southeast, its coastal sands and related deposits covered canyon country entirely.

The sea eventually stopped its westward journey as it encountered the mountainous highlands that were first thrust up in central Utah during the Jurassic period, but continued northward adjacent to the highlands. These highlands rose still higher during the Cretaceous, and provided great quantities of sediments for the canyon country region.

This great Cretaceous sea was the product of a major continental drift event which caused a worldwide rise in the sea level of about 1500 feet. This inundated some forty percent of the Earth's land surface at its maximum, including all of canyon country. The sea encroached from the southeast through a narrow strait in the New Mexico-Texas border area and expanded into New Mexico, Colorado and Utah. It next continued northward, blocked on the west by the highlands that began in southern Nevada and central Utah and extended north-westward clear to Alaska. This arm of the southern ocean eventually joined an arm of the Arctic ocean reaching south through Canada, Montana and Wyoming, but apparently this connection also failed to provide much circulation for the inland sea. The restricted connections between this immense inland sea and the oceans to the north and southeast created relative stagnation within the sea, a factor that later contributed to the present shape and color of canyon country.

As this great sea settled over the region, its level rose and fell gradually over a long period of time, causing its shoreline to advance and recede and producing an interlayering of land, coastal and marine deposits in various areas within or bordering canyon country. As the sea lay across canyon country, the western part of the region slowly sank, allowing these deposits to accumulate there to great depths. Much of the material for the coastal and marine deposits of this period came from the range of highlands to the west.

The continental drift event that triggered the invasion of the Cretaceous sea finally ended in the latter part of this period. Then the sea began its slow retreat out of canyon country. As it receded, with minor advances at times, the marine deposits it left behind were covered by sediments from the western highlands, until finally all of canyon country was buried beneath them.

During this slow retreat of the sea, its coastlands were sandy beaches, behind which were heavily vegetated marshes, lagoons, and low, flat coastal plains. Dinosaurs occupied these areas, but as the Cretaceous sea slowly retreated, the age of giant reptiles came to a close all over the world. As the rich coastal swamps were buried beneath other sediments, the organic matter they contained was slowly converted into the coal now found where these late-Cretaceous deposits remain in and near canyon country.

A final intrusion of the sea barely entered the eastern edge of canyon country, but a period of erosion early in the Tertiary period removed most of these final deposits within the region.

It should be noted that the great Cretaceous sea was the last marine invasion of the canyon country region. From that time until the present, canyon country was no longer covered by seas, nor was it even coastal. It became the inland region it now is. All regional deposits subsequent to the Cretaceous period are either fresh water sediments or igneous in nature.

Toward the latter part of the Cretaceous period, a series of major events began that were to play a major part in shaping canyon country. Eastward pressures against the west coast of the North American continent, started by a major continental drift event, had begun in the late Jurassic and slowly built up, further uplifting the Sierra Nevada range of mountains, then moving on eastward into western Utah, fracturing the continental crust as it traveled and initiating an interval of volcanism over the entire western region. In western Utah, the earlier uplift was given the additional boost noted in the first paragraph of the Cretaceous history.

When this eastward-traveling crustal pressure reached the much thicker Colorado Plateau province, it triggered a series of violent but slow-motion events within and around the plateau. This activity continued on into the Tertiary period and produced many of the geologic features most prominent in canyon country today.

Within canyon country, as defined for the purposes of this book, the slow but gigantic crustal pressures from the west coast produced several major and minor uplifts, including the San Rafael Swell and the Monument and Circle Cliffs uplifts. In the regions bordering canyon country, the Uncompahgre uplift was reactivated, the Uinta, San Juan, Zuni, Defiance and Kaibab uplifts were produced and the Uinta, San Juan, Black Mesa and Kaiparowitz basins were formed. Farther east, in a wide belt that ran from New Mexico up into Canada, the same pressures gave rise to the present Rocky Mountains. The formation of this range continued on into the Tertiary period, and subsequent glaciation helped shape it further.

Crustal pressure on the Colorado Plateau also produced a ring of volcanism around the plateau that began in the late Cretaceous and continued on into the next period, the Tertiary. Evidently, the plateau's thicker crust did not permit volcanoes within its boundaries, but did eventually allow the formation of several small isolated "laccolithic" mountain ranges within the plateau. These will be described more fully in the Tertiary history and in later chapters under other headings.

Extensive erosion during the periods that followed the Cretaceous removed vast quantities of Cretaceous deposits from canyon country, especially during the last 10 million years. Even so, Cretaceous rock can be seen in many places within the region, and the effects of the crustal activity that started in the late Cretaceous are spectacularly obvious as mountain ranges and various highlands.

Tertiary (1 to 65 MY). The major crustal disturbances that began to affect canyon country in the late Cretaceous period continued on into the Tertiary. Sediments from the new highlands gradually accumulated in the lowlands and new basins. The entire general region was still at a fairly low elevation above sea level, and the mountain ranges within and adjacent to the region did not stand out in sharp contrast as they do today. In the mid-Tertiary, the slowing interval of crustal deformation phased into a time of volcanism in the regions surrounding canyon country.

A major uplift involving several western states occurred in the late Tertiary. Changes in climate and erosion rates brought about by this uplift removed thousands of feet of Cretaceous and Tertiary deposits from canyon country, and were largely responsible for shaping the region as it is now seen, with Tertiary sedimentary rock left only in the Book Cliffs area to the north. This land-shaping process is continuing even now, at a fairly rapid rate as geologic   processes go.

Many events of importance to the present canyon country scene happened during the Tertiary period. Because of this, more detailed descriptions of this period are made by epochs within the Tertiary in the following pages. Figure 4, on the back cover, illustrates these subdivisions on the geologic time scale.

Tertiary/Paleocene (53 to 65 MY). The uplift and basin-forming crustal activities begun earlier continued through this epoch. Within canyon country, water-borne sediments from the various new highlands within and adjacent to the region gradually filled the lowlands and newly forming basins. A huge freshwater lake to the west accumulated the colorful sediments that are now exposed in Bryce Canyon National Park, Cedar Breaks National Monument and vicinity.

At about the same time, stream deposits from the areas surrounding canyon country, principally from the east, were building up in the northern part of the region and interlayering with the lake deposits to the west. This lake-and-stream interval continued on into the next epoch, the Eocene.

Tertiary/Eocene (39 to 53 MY). The crustal deformation activities started earlier continued through most of this epoch, but tapered off toward its final years. Erosion continued the leveling process. Early in the Eocene, uplift and freshwater sedimentation patterns created another immense lake to the north of canyon country. Deposits from this lake extended into the northern edge of canyon country and interlayered with stream-borne material from the surrounding highlands. The lake basin slowly sank during this epoch, and thus accumulated thick deposits during its life.

Sedimentary rock strata from this Eocene lake are the youngest now remaining in canyon country, and are visible only as the uppermost strata of the Book Cliffs in the north of the region, although younger rock of an igneous nature can be found in several areas.

Toward the end of this epoch, a new and isolated range, the Henry Mountains, appeared in the center of canyon country. Hot igneous rock flowed upward into the overlying sedimentary rock layers through several vertical columns, and bulged the rock layers upward still further by flowing outward between them in many places, and upward into vertical cracks in the rock. Since this flow of lava-like rock did not break through on the surface, it is called "intrusive." Had the hot, flowing rock broken the surface, it would have been called "extrusive" and the mountains it formed would be volcanoes. Navajo Mountain may have formed about this time too, as a single, isolated intrusive dome.

About this time, the same forces initiated volcanic activity all around the Colorado Plateau. The Plateau's thicker crust prevented volcanism within canyon country, but did permit the formation of three non-volcanic mountain ranges, the one just noted, plus two others later. Some volcanic activity around the perimeter of the Plateau has continued into recent times, particularly to the southwest, in central Arizona.

Black igneous rock in Capitol Reef National Park.

Tertiary/Oligocene (25 to 39 MY). During this epoch, the volcanic activity all around the Colorado Plateau continued and spread, covering vast regions with lava, ash and other types of igneous rock. Some of this extensive flow even entered canyon country in the vicinity of Capitol Reef National Park. The Uncompahgre uplifted for the sixth time early in this epoch.

While all this volcanic activity was going on around canyon country, erosion within the region continued to wear down the highlands, but at a very slow pace. The entire region, including the upland to the east, was rather arid, with gentle contours and few elevation extremes. These factors combined to place limits on water erosion processes. The present-day system of rivers and perennial streams had not yet developed. Most water courses were intermittent, and thus sediments from higher ground tended simply to accumulate in the lower areas, rather than be carried from the region by running water.

Toward the end of this epoch, two more isolated mountain ranges, the Abajos and La Sals, formed in eastern canyon country by intrusive igneous activity. These two ranges, together with the Henrys and Navajo Mountain that formed in the center of the region during the mid-Eocene, are now prominent geologic features in canyon country.

Tertiary/Miocene (11 to 25 MY). The volcanism that began around the perimeter of the Colorado Plateau and continued through the Oligocene, dwindled to a low level in the early Miocene, because the continental drift event that had caused it ended.

The ending of another continental drift event a little later also released eastward crustal pressures. This allowed a slow stretching out of the massively faulted region of western Utah, Nevada and parts of Arizona and California. The stretching extended this region as much as sixty miles, and formed the extensive basin-and-range province there now, but had little effect upon canyon country. Again, the much thicker crust of the Colorado Plateau protected canyon country from forces that drastically changed nearby regions with thinner crusts.

Erosional processes continued at their slow pace through this epoch, much as they had during the previous epoch, and for the same reasons, although the formation of the two intrusive mountain ranges in eastern Utah did increase erosion rates in their vicinities.

Tertiary/Pliocene (1 to 11 MY). As the Miocene epoch phased into the Pliocene, the last Tertiary epoch, a major event began which was to contribute, more than any other single event, to the shaping of canyon country as it is today.

For reasons still not determined, but probably related to west coast continental drift activities, the entire general region east of the Sierra Nevada mountains, including the Rocky Mountain range, began to rise. This immense uplift eventually left the entire region more than a mile higher than it had been, and changed its topography drastically. This regional uplift also triggered a seventh rise of the Uncompahgre

Uplift early in the Pliocene epoch.

As the region's elevation became higher and higher, its climate became wetter and wetter, especially in the mountains and other highlands. Precipitation increased slowly but steadily, which in turn created a system of perennial meandering rivers and streams throughout canyon country and adjacent Colorado Plateau and mountainous regions. As the uplift continued, still more water fell on the region, increasing erosion rates and removing the most recent deposits at a rapid rate. The water courses cut deeper and deeper, shaping canyon country as it is today.

Toward the end of this epoch, and perhaps on into the early Quaternary period, the Uncompahgre Uplift to the east of canyon country rose still higher for the eighth time, adding further impetus to the quickening erosion of the region.

Quaternary/Pleistocene (10,000 to 1 MY). The general regional uplift that began early in the Tertiary/Pliocene epoch, and which accelerated erosion within canyon country, also initiated an interval of lake-forming in the regions surrounding canyon country, and contributed to the glaciation that took place during the Quaternary period.

In western Utah, Lake Bonneville formed, reaching a depth of 1000 feet and covering some 20,000 square miles at its greatest extent. Highlands in central Utah kept this immense lake out of canyon country.

During the Quaternary/Pleistocene epoch, an ice age covered the northern part of the North American continent, but the continental glaciers did not reach canyon country. At the same time, however, separate glaciers formed in canyon country's three mountain ranges, the Abajos, Henrys and La Sals, at elevations above 10,000 feet, and on bordering uplifts. Navajo Mountain was not high enough for glaciation.

While glaciers were shaping some of the higher canyon country peaks and passes, accelerated erosion processes continued to carry away cubic miles of Cretaceous and Tertiary deposits and to expand and deepen the vast canyon system being carved into older strata by the Colorado River and its many tributaries. Increased water runoff during intervals of glacier melting and retreat accelerated erosion in major canyon country river gorges such as the Colorado, Green, Dolores and San Juan, and in some of their tributaries.

The general regional uplift which started early in the Tertiary/Pliocene initiated another unusual geologic process which continues even now. The salt flow domes and ridges that began to form in canyon country during the Permian period were gradually exposed to ground and surface waters which dissolved the more soluble parts of the salt masses. As the region's rivers cut deeper, the dissolved salt could drain away, allowing rock strata overlying the remaining salt to sink and thus slowly form the present "salt valley" features so prominent in the vicinity of the La Sal Mountains.

Navajo Mountain.

Quaternary/Holocene (present to 10,000 years ago). The various erosion processes that were quickened by the last major regional uplift continued during this epoch, and are still at work shaping canyon country today. The advent of man in canyon country within the last 100 years has further accelerated these erosional processes almost to a catastrophic level, as geologic processes go.

Another period of glaciation during this final epoch of canyon country's geologic history again brought glaciers to the La Sals, the highest of the three canyon country ranges, during a brief glacial interval that lasted from 1200 to 1850 A.D. Glacial moraine deposits are plentiful in some areas of the La Sal Mountains, and around Boulder Mountain, in western canyon country.

SUMMARY

From the ancient Precambrian era to the late Cretaceous period, about 80 million years ago, the canyon country region was essentially low and relatively flat. It was either under shallow seas, or was coastal or low coastal plains. During this time, it accumulated many layers of marine, coastal and fresh water deposits, and its climate varied from tropical or subtropical to temperate or arid, as desert-arid as the Sahara Desert is today.

The fluctuations in sea level and climate were largely caused by various continental drift events taking place during that time. Because the continental drift concept is so new, many of the relationships between continental drift events and other geologic history events have not yet been determined, but are still being studied.

Within this same span of time, the only major feature to develop above the general land level in canyon country was the Uncompahgre Uplift, which started rising about 310 million years ago and continued by stages until very recently. This uplift contributed toward the composition and shaping of canyon country in many ways.

From the late Cretaceous to the late Tertiary periods, an interval of about 55 million years, canyon country and vicinity was subjected to violent crustal activity and deformation. This produced a variety of faults, uplifts and mountain ranges within canyon country, raised existing uplifts on the borderlands still further, and created whole new ranges of mountains, the Uintas and Rockies, immediately to the north and east.

Various deformation events during this time were caused by continental drift events. Some of these relationships are fairly well understood. Others are not, but research continues.

Despite the new uplifts and basins formed within canyon country, and the more rapid erosion rates in their vicinity, the canyon country region was still low in elevation, gently contoured and too arid for much shaping by water erosion. The present complex system of rivers and streams did not yet exist.

In the late Tertiary about 10 million years ago, the entire canyon country region was slowly uplifted about one mile. The changed climate and increased water runoff created the Colorado River and its many tributaries, and began carving canyon country into its present shape. This immense uplift, which affected much of the western United States, was undoubtedly caused by some continental drift event or phenomenon. Exactly what event has yet to be determined, but one hypothesis is noted in a later chapter on major shaping events.

As the accelerated erosion processes caused by the regional uplift continued up to the present, the higher mountain peaks of the three canyon country mountain ranges were lightly shaped by glaciers during the last ice age that ended about 10,000 years ago. Then the La Sals received a final glacial touch within the last thousand years.

At present, all geologic processes within canyon country, or that could affect the region, are fairly stable, so far as geologic scientists know, although human activities are affecting erosional rates in several direct and indirect ways.

Cutler Formation, White Rim, Canyonlands National Park.

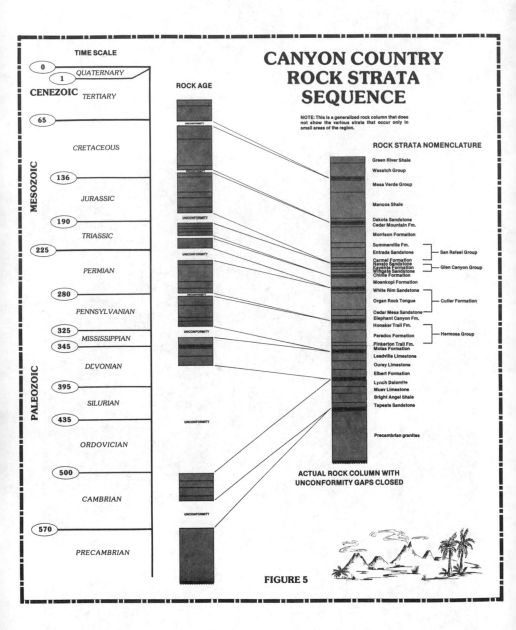

CANYON COUNTRY ROCK STRATA SEQUENCE

TIME SCALE

CENEZOIC
- QUATERNARY
- TERTIARY

0
1
65

MESOZOIC
- CRETACEOUS
- JURASSIC
- TRIASSIC

136
190
225

PALEOZOIC
- PERMIAN
- PENNSYLVANIAN
- MISSISSIPPIAN
- DEVONIAN
- SILURIAN
- ORDOVICIAN
- CAMBRIAN

280
325
345
395
435
500
570

PRECAMBRIAN

ROCK AGE

NOTE: This is a generalized rock column that does not show the various strata that occur only in small areas of the region.

ROCK STRATA NOMENCLATURE

- Green River Shale
- Wasatch Group
- Mesa Verde Group
- Mancos Shale
- Dakota Sandstone
- Cedar Mountain Fm.
- Morrison Formation
- Summerville Fm.
- Entrada Sandstone — San Rafael Group
- Carmel Formation
- Navajo Sandstone
- Kayenta Formation — Glen Canyon Group
- Wingate Sandstone
- Chinle Formation
- Moenkopi Formation
- White Rim Sandstone
- Organ Rock Tongue — Cutler Formation
- Cedar Mesa Sandstone
- Elephant Canyon Fm.
- Honaker Trail Fm.
- Paradox Formation — Hermosa Group
- Pinkerton Trail Fm.
- Molas Formation
- Leadville Limestone
- Ouray Limestone
- Elbert Formation
- Lynch Dolomite
- Muav Limestone
- Bright Angel Shale
- Tapeats Sandstone
- Precambrian granites

UNCONFORMITY (repeated at various levels)

ACTUAL ROCK COLUMN WITH UNCONFORMITY GAPS CLOSED

FIGURE 5

34

ROCK STRATA

GENERAL

As many kinds of sediments were deposited during the various periods of geologic history, each new layer buried the next older layer, usually after mixing somewhat with its upper surface. As the older layers were covered, they slowly turned to stone of various types. Sandy strata, such as desert dunes, beach sand, offshore sand bars and some types of sandy freshwater sediments became sandstone. Finer sediments became mudstone and siltstone. Sediments mixed with gravel and broken rock became conglomerate. Layers of salts that were either precipitated out of water, or left when it evaporated, became limestones, or crystallized salt strata.

Sand and courser materials were bound together into solid rock by chemical cements brought by seeping water. Finer sediments were also held together by interlocking clay particles. Pressure from the weight of overlying strata and water aided the petrifying process.

Thus, throughout geologic history, sedimentary layers accumulated and were slowly turned to rock of various types, with each layer being older than the one above it and younger than the one below. There are two exceptions to this. Whenever rock layers are so severely folded by later geologic events that their sequence is inverted, within a limited area older strata may occur above younger ones. A second exception is when igneous activity forces hot lava-like rock between layers of sedimentary rock. The resulting igneous rock layer is obviously younger than the strata above it.

Another general rule about sedimentary rock layers is that they are laid down very nearly horizontal. Rock strata that are not now horizontal, or which have become misshaped in some manner, have been changed by later geological events. Some of the major events that have shaped canyon country rock strata are discussed in a later chapter.

Sometimes sedimentary rock is changed into another form and chemical composition by heat and pressure deep beneath the earth's surface, or in the vicinity of igneous activity. Changed chemical environments can also convert rock from one form to another. Some kinds of very old, deeply buried sedimentary rock turn into granite-type rock of one kind or another. Sedimentary rock changed by later conditions is called "metamorphic" rock.

The study of rock layers is a very important tool in geology, and is one key to understanding geologic history and processes. The thickness and composition of rock strata, their sequence and other less obvious factors, all contribute toward an understanding of the earth's past. A time gap, or unconformity, in the rock column may also contribute knowledge, although the detection of these gaps in the stratigraphic record may sometimes be difficult.

The sequence of rock strata is generally determined by drilling. The cores of such drillings are then studied and compared. Modern echo-sounding techniques also provide information about subsurface rock strata.

In canyon country, deeply cut canyons and river gorges also expose thousands of feet of rock strata. Geologists study all of this information and from it slowly build an overall understanding of the geologic history of an area, a region, a continent and the entire world.

Figure 5 shows how canyon country rock strata fit onto the geologic time scale. In one stratigraphic column the major unconformities, or gaps in the rock history due to erosion that were noted in the previous chapter, are shown as gaps in the rock sequence. In the other column, these gaps are closed, and indicated only by wavy lines, and the rock strata appear just as they do now in canyon country.

It should be noted that these two stratigraphic columns are generalized for the canyon country region. Variations in rock layers occur from area to area. Some of these variations are discussed when the individual rock strata are described in the following pages.

It could be said that Figure 5 is a "calendar" of canyon country geologic history. In the following pages, each "date" or layer in this calendar is described in more detail, and related to the general geologic history summarized in the previous chapter.

Geologists use standardized nomenclature in describing various rock strata. Each layer is assigned a name, generally after the geographic locality in which it was first studied, plus a term which describes the type of rock it is. If a rock layer varies somewhat in composition from place to place, or consists of several different subunits of rock, it is called a "formation." If the layer is of fairly uniform composition, it is labeled with the appropriate term, such as "sandstone," "limestone," "shale" or "conglomerate."

Several "formations" that seem to be related to each other in some way, such as having a common source, are sometimes called a "group." Distinctive rock layers within a formation are called "members." A rock member with limited extent may be called a "tongue." Some examples: the Moab Tongue of Entrada Sandstone; the Cedar Mesa Sandstone member of the Cutler Formation; and Kayenta Formation within the Glen Canyon Group. Most of the distinctions implied by such nomenclature are of interest only to geologists, but a general familiarity with the noted terms will be helpful in understanding the following descriptions of individual canyon country rock strata.

The rock strata nomenclature used in these descriptions is as standardized for the canyon country region by the Four Corners Geological Society during its Eighth Field Conference in September 1975.

Rock strata are listed from the oldest to the youngest, with major gaps in the rock record noted as unconformities. Most such unconformities represent intervals of erosion, during which more sediments were removed than deposited.

STRATA DETAILS
Precambrian Era.

The Precambrian rock beneath canyon country is between 570 and 2,000 million years old. It is largely marine sediments that have been changed into metamorphic granite-type rock. Formal names have not been assigned to Precambrian rock in this region.

Precambrian rock is generally deep underground in most of canyon country, but is exposed in and near the Colorado River gorge in Westwater Canyon, beginning about 6 miles west of the Colorado-Utah boundary, and in Unaweep Canyon along Colorado 141, to the northeast of Gateway, Colorado, where the ancient spine of the Uncompahgre Uplift is exposed.

Cambrian Period

Unconformity. A long interval of erosion removed deposits from the first part of the Cambrian period.

Ignacio Quartzite: quartzite and sandy shale, of marine origin.

Bright Angel Shale: shale and sandstone, of marine origin.

Muav Limestone: shale and limestone, of marine origin.

Lynch Dolomite: dolomite, a chemically modified limestone, of marine origin.

All of these late Cambrian sedimentary rock strata get thinner and younger from west to east, as the Cambrian sea slowly invaded in that direction. Cambrian rocks occur under most of canyon country, but are not exposed on the surface.

Unconformity. An interval of erosion removed any late Cambrian deposits.

Ordovician and Silurian Periods.
A long interval of erosion removed any deposits from these two periods. Rocks representing these periods do not occur under canyon country.

Devonian Period.

Unconformity. The interval of erosion that removed all deposits from the two previous periods continued well into the Devonian period.

Ouray Limestone: limestone, with a little shale, of marine origin.

Unconformity. There was a short interval of erosion at the end of the Devonian period.

Mississippian Period.

Leadville Limestone: limestone and dolomite, of marine origin.

Unconformity. An interval of erosion removed all deposits from the last half of this period.
Mississippian rock occurs under most of canyon country, but is rarely found exposed on the surface. Rocks of this period have been identified in Salt Valley, in Arches National Park, but the origin of these isolated boulders is not known.

Pennsylvanian Period.

Unconformity. The interval of erosion that ended the Mississippian period, continued briefly into the Pennsylvanian period.

Molas Formation: shale, sandstone, mudstone and thin limestone, of mixed marine and land origin.

Pinkerton Trail Formation: limestone, shale, sandstone and siltstone, of marine origin.

Paradox Formation: primarily crystalline salts, with added layers of shale and dolomite of restricted marine origin. See later description of "salt valleys" for details. Original thickness up to 5000 feet, but may be up to 14,000 feet thick under certain salt valleys. Generally occurs underground, but exposures can be seen in Fisher Valley, Moab Valley near the river portal, lower Salt Valley in Arches National Park and in other salt valleys to the south and east of the La Sal Mountains.
The crystalline salts in this formation precipitated from relatively

Paradox Formation salts, Fisher Valley.

stagnant, evaporating sea water that was renewed many times. Where exposed, it is rain-eroded into dirty white, sharp-crested hillocks. It is usually banded with various colors, and its crystalline structure is quite apparent. Thin layers of similar material occur within other canyon country rock formations, but only beneath the huge Paradox Basin are such salt deposits so extensive and deep.

Honaker Trail Formation: interlayered shale, limestone and sandstone, of mixed marine and land origin. This formation was laid down as a sea retreated from the canyon country region, hence the mixture of land and marine sediments. The deposits came primarily from the Uncompahgre Uplift highlands to the northeast of the region, and the formation is up to 3000 feet thick.

This formation is the oldest sedimentary rock layer that is exposed to view to any extent within canyon country. It is exposed within and adjacent to the Colorado River gorge beginning just beyond the end of Utah 279 and continuing downriver almost to Hite. There is another large exposure in the San Juan River gorge downriver from Mexican Hat, and a small one to the west of U.S. 163, opposite the visitor center at the entrance to Arches National Park.

The part of this formation that represents the final transition zone between Honaker Trail marine sediments and the following land sediments is sometimes called the "Rico Formation." Three off-road vehicle trails, named Potash, Jackson Hole and Chicken Corners, travel on or near a prominent fossiliferous limestone layer within the "Rico Formation" downriver from Moab. For more details, see the later chapter on rockhounding. The Lockhart Canyon spur of the Lockhart Basin trail permits off-road vehicles to penetrate a canyon walled by the upper part of the Honaker Trail Formation.

The Honaker Trail Formation varies in color. Different layers may be red, white, gray or combinations of these hues. The formation erodes into near-vertical cliffs, with harder layers forming narrow shelves or terraces. The prominent gray-colored limestone layer just noted is somewhat harder and serves to protect lower strata from surface erosion along the river gorge downriver from Moab.

The Pinkerton Trail, Paradox and Honaker Trail formations are collectively called the Hermosa Group.

Unconformity. An interval of erosion removed some late Pennsylvanian deposits. Very little was removed in eastern canyon country, but the Honaker Trail Formation gets thinner toward the west, and most or all Pennsylvanian deposits were eroded away in the San Rafael Swell vicinity.

Permian Period.

Elephant Canyon Formation: limestone and dolomite interlayered with sandstone, siltstone and thin beds of crystalline salts, of marine origin. This formation was deposited by the late Pennsylvanian sea that lingered in northwestern canyon country, and ranges up to 1500 feet thick in the vicinity of Canyonland National Park. In the southern part of the region, deposits left on coastal plains beside this sea are called Halgaito Shale. These red sandstone and sandy shale deposits reach depths of 500 feet, and are sometimes classified as a part of the overlying Cutler Formation. Elephant Canyon and Halgaito sediments came largely from the Uncompahgre Uplift to the northeast, with Elephant Canyon deposits modified by the marine environment.

In appearance, the Elephant Canyon Formation somewhat resembles the earlier Honaker Trail Formation, being intermixed layers of varying hardness and color, with red dominating. Exposures of this formation are not readily available except to hikers who go down to the Colorado River via the various trails and routes in the Needles District of Canyonlands National Park. The reddish Halgaito Shale can be seen in the exposed strata of the anticline visible from U.S. 163 in the vicinity of Mexican Hat, and at the bases of the spires and buttes in Valley of the Gods to the north of that town.

Cutler Formation: sandstone, conglomerate, sandy shale and limestone, with the freshwater sediments interlayered with the dune sand and marine deposits. The land sediments and sand came primarily from the Uncompahgre highlands to the north and east and are several thousand feet thick near the uplift. In the region between this uplift and Moab and Monticello, the Cutler Formation is relatively uniform. To the north, west and south, four unique tongues or members of the Cutler dominate.

Unseparated Cutler Formation is exposed many places in eastern canyon country. U.S. 163 travels through dark red Cutler deposits beginning about ½ mile north of the entrance to Arches National Park and continuing on to the U.S. 163/Utah 313 junction. To the east of Moab, the Onion Creek trail twists for miles through a maze of eroded Cutler sandstone. The harder Cutler strata erode into strangely rounded walls, columns and shapes.

Cutler/Cedar Mesa Sandstone member: sandstone, of coastal-marine origin, up to 1200 feet thick. This light-colored rock formed from beach and sandbar sand that was left by a sea invading from the northwest. The sea stopped before reaching as far east as U.S. 163, but inter-layered with red Uncompahgre sediments along its eastern shores, to produce the red-and-white banded sandstone prominent in the Needles District of Canyonlands National Park.

This sandstone erodes to form vertical cliffs and rounded fins and domes. It also forms caves, arches and natural bridges quite easily. Cedar Mesa Sandstone is exposed on the surface over vast areas of canyon country, primarily where exposed by erosion on the upper slopes of the immense Monument Uplift that begins in the Needles District and continues south into Arizona. Some other outstanding exposures of this beautiful sandstone are in the Maze District of Canyonlands National Park and the giant spans in Natural Bridges National Monument.

Cutler/Organ Rock tongue: sandstone, siltstone and shale, of land origin. These red fresh-water sediments accumulated in tidal flats that followed the retreat of the sea that deposited Cedar Mesa Sandstone. Organ Rock deposits are up to 900 feet thick, and are found under or on the surface of most of the canyon country region except the north and northeastern parts.

Organ Rock erodes to form red slopes with ledges of harder sandstone. It is exposed in many places across the Monument Uplift, but is clearly visible near Utah 95 as the red layer just above the dominant white Cedar Mesa Sandstone, between the Natural Bridges vicinity and Hite on upper Lake Powell. It also forms the lower slopes below the towering spires of Monument Valley, near where U.S. 163 crosses the Utah-Arizona border.

White Rim Sandstone, Canyonlands National Park.

Cutler/White Rim Sandstone: white sandstone, of coastal-marine origin. This distinctive rock layer is up to 250 feet thick. It consists of beach and offshore sand left by a second sea invasion during this period that penetrated canyon country from the northwest, but only as far east as the present Colorado River.

At its edge, White Rim Sandstone erodes into vertical or undercut cliffs. Detached chunks, supported by pedestals of softer underlying rock, form immense balanced rocks, and dark columns protected by huge, white caprocks. On its upper surface, White Rim Sandstone is quite resistent to water, but erodes into shallow depressions and low mounds, and is often scored in regular patterns by vertical cracks and joints that penetrate the rock for its full depth. It is exposed almost exclusively along the White Rim, near the White Rim trail, in the Island in the Sky District of Canyonlands National Park, but is also visible as a white band high in the red bluffs above the Hite area of Lake Powell, near Utah 95, and a few other less accessible places.

Beyond the western extent of the Organ Rock tongue, the Cedar Mesa and White Rim sandstones are not separated by the Organ Rock tongue and together reach a maximum of 1200 feet thick. This combination is sometimes erroneously called "Coconino Sandstone."

Cutler/deChelly Sandstone: sandstone, primarily of a desert-dune origin. This salmon-hued sandstone formed from vast expanses of desert sand dunes that covered the southern canyon country region at about the same time as the White Rim beach sands were being deposited farther north. It is 1000 feet thick at its maximum in northern Arizona.

DeChelly Sandstone erodes into vertical cliffs or spires, with rounded domes and fins on its upper surface. It is best known for forming the spectacular spires and buttes of Monument Valley, and the magnificent canyon system in Canyon de Chelly National Monument.

Kaibab Limestone: gray limestone, of marine origin. This limestone was deposited by a late Permian sea that barely entered the western fringes of canyon country in the vicinity of the San Rafael Swell and Bryce Canyon National Park. Kaibab Limestone is exposed many places in the eastern canyons of the San Rafael Swell, but one exposure is traversed by Interstate 70 about 8 miles west of the I-70/Utah 24 junction.

Unconformity. An interval of erosion removed some late Permian deposits, with more removed from the higher ground in eastern canyon country.

Triassic Period.

Unconformity. The interval of erosion that removed some late Permian deposits also removed some early Triassic deposits.

Moenkopi Formation: shale, siltstone, sandstone and limestone, of interlayered shallow marine, tideland and mudflat origin. Deposits are predominantly red and brown, with layers of gray limestone toward the western part of the region. When exposed to weathering, the harder layers of the formation erode into strangely convoluted walls, columns and shapes, while the softer shales form gentle slopes ledged by thin harder layers.

Geologists separate this formation into several different members in various areas of canyon country, but such distinctions have little meaning to non-professionals. The whole formation is 800 feet thick at its maximum in most places, but exceeds 2000 feet thick in some sunken areas adjacent to salt valleys and in some areas to the west of the canyon country region.

The Moenkopi Formation is exposed throughout canyon country, with huge areas of the reddish rock along the flanks of the ancient Monument Uplift, in the San Rafael Swell and to the west of the Waterpocket Fold. Utah 24 goes through one beautiful area between the visitor center in Capital Reef National Park and Torrey to the west. The red slopes above the White Rim off-road vehicle trail in Canyonlands National Park are Moenkopi, as are the red, sloping hills in lower Castle and Professor valleys along Utah 128, upriver of Moab.

Moenkopi Formation, Chimney Rock, Capitol Reef National Park.

Unconformity. An interval of erosion removed considerable mid-Triassic deposits from canyon country.

Chinle Formation: shale, somewhat sandy and silty and with localized layers of limestone conglomerate, of land origin, with a liberal admixture of airborne volcanic ash from Arizona. Chinle deposits were laid down on vast, level flood-plains by meandering streams and shallow lakes, using material carried by freshwater flow from the low-profile remnants of the Uncompahgre Uplift and Ancestral Rockies to the north and east. Chinle sediments eventually covered the exposed Precambrian rock spine of the Uncompahgre, and this relationship is now visible on the re-exposed uplift.

The Chinle Formation varies up to 1200 feet thick and is exposed many places within canyon country. It erodes into fairly steep slopes set with ledges of harder rock, and is variegated in color, with gray dominating in the layers that have been impregnated with volcanic ash. When unprotected by harder rock above it, the Chinle weathers into low, rounded hills.

As with the Moenkopi Formation, geologists have subdivided the Chinle Formation into several members, primarily in the western areas of canyon country. These members do not extend into eastern canyon country, but represent localized conditions in the western part. Most of the distinctions between various Chinle Formation members are of interest only to professional geologists, although the various members do exhibit a variety of colors and manners of erosion, and thus add to the scenic beauty of the region.

Chinle deposits are the easiest to identify as the first grayish layers below the high vertical Wingate Sandstone cliffs that are so outstanding in the eastern half of canyon country. The Chinle Formation is being mined for uranium just west of U.S. 163, for a mile or so north of the U.S. 163/Utah 313 junction. It is also visible as the gray, sloping base for the soaring cliffs of Dead Horse Point State Park and the Island in the Sky District of Canyonlands National Park. In Capitol Reef National Park, the colorful slopes immediately below the soaring Wingate cliffs to the east of the scenic drive south from the campground are also Chinle deposits.

More open exposures of the Chinle Formation occur around the lower slopes of the Monument Uplift, in the Circle Cliffs area to the west of Waterpocket Fold and in other areas on south into Arizona. Most open areas of Chinle require off-road vehicles for access, but a huge exposure is available to Lake Powell boaters along the San Juan River arm of the lake, in the Paiute Creek-Monitor Butte vicinity.

Wingate Sandstone: red-tinted sandstone, of desert-dune origin, with some interlayering of desert dry-lake sediments. This sandstone was deposited when a vast, Sahara-like desert covered canyon country and adjacent regions. It is up to 400 feet thick, and erodes into vertical-walled cliffs, peninsulas and spires. Its upper surface is generally

45

Wingate Sandstone cliffs, The Palisade, Gateway, Colorado.

protected by overlying Kayenta Sandstone, but when it is exposed, it weathers into rounded domes and fins, and occasionally forms arches and natural bridges.

Wingate Sandstone is exposed to view many places throughout canyon country as vertical cliffs. All such cliffs in the Island in the Sky District of Canyonlands National Park, in Dead Horse Point State Park and around the base of immense Hatch Point of the Canyon Rims Recreation Area are Wingate, as are many of the sheer cliffs in Capitol Reef National Park on the west side of Waterpocket Fold.

Wingate cliffs can be confused with similar sheer cliffs of other sandstones, such as Cedar Mesa, White Rim, Navajo and Entrada, but positive identification is made easy by noting the distinctive rock formations immediately above and below Wingate cliffs. This sure and easy identification makes Wingate Sandstone one of the "keys" to identification of all other strata exposed above and below it.

In northern Arizona, the Moenave Formation, a fresh water floodplain deposit, replaces the upper part of the Wingate Sandstone.

The massive, sheer cliffs of Wingate Sandstone form a prominent feature in canyon country, and add immeasurably to its majesty and beauty.

Kayenta Formation: Sandstone and shale, of freshwater sedimentary origin. This grayish rock, with red or purple tints in places, was deposited in alternating hard and soft layers of varying thickness by a system of sluggish streams that covered most of canyon country following the Wingate desert interval. It varies in thickness up to 300 feet or more.

The upper surface of the Kayenta Formation erodes into expanses and slabs of harder sandstone, undercut and tilted wherever the softer underlying layers have been reached by water erosion. At cliff edges, where the Kayenta is commonly exposed above cliffs of Wingate Sandstone, harder Kayenta layers form ledges and terraces, with their edges collapsing as the softer layers erode away. Open expanses of the upper surface of the Kayenta Formation are fairly common and are typified by relatively flat terrain partially covered by sand dunes, set with occasional remaining domes of overlying Navajo Sandstone, and with the uppermost Kayenta layers cracked and tilted by water erosion.

The Gemini Bridges off-road vehicle trail near Moab goes through a typical open Kayenta exposure, and elsewhere throughout much of canyon country, Wingate Sandstone cliffs are topped by the characteristic horizontal layers of the Kayenta Formation.

Navajo Sandstone: white sandstone, of desert-dune origin, but with occasional tints of red, yellow, brown or other colors. Thickness varies up to 600 feet. Returning desert dunes, set with occasional dry lakes, formed this distinctive canyon country rock. In some nearby regions, the Kayenta water-borne deposits did not separate the Wingate and Navajo sand dune deposits, and this combination bears the name "Nugget Sandstone." The combination of Wingate, Kayenta and Navajo was named the Glen Canyon Group, because it dominates that region of the Colorado River gorge, now Lake Powell.

Navajo Sandstone strength, texture and cohesion are such that the rock cannot support its own weight except under compression, as in arches or arched alcoves. Because of this, when it is undercut, or loses support from underlying Kayenta layers, it collapses in a near-vertical plane, much the same as other massive sandstones do. Thus, along water courses or cliff edges, Navajo Sandstone erodes into sheer cliffs. When its upper surface is exposed, as it is quite widely throughout canyon country, it erodes into a complex of domes, fins and rounded surfaces, often along patterns set by earlier cracks and joints. Where the weathered rock still has a low profile, it often resembles an expanse of desert sand dunes, frozen into immobile rock by some mysterious geologic process.

There are vast expanses of exposed Navajo Sandstone "petrified dunes" in canyon country. The paved road into Arches National Park provides views to the southeast across one such area, and the Sand Flats, Poison Spider Mesa and Moab Rim off-road vehicle trails traverse others. The entire highlands in all directions from Lake

Navajo Sandstone cliffs, Colorado River gorge.

Powell, between Forbidden Canyon and Red Canyon far up the lake, are almost solid Navajo Sandstone, as is the entire area surrounding the Escalante Canyon system. Navajo also makes up the highest rock along much of the great Waterpocket Fold, from upper Capitol Reef National Park to Lake Powell. It also forms the jagged eastern "reef" of the San Rafael Swell, and the spectacular cliffs of Zion National Park.

Navajo Sandstone, because of its weathering characteristics, also forms spring-seep caves, arches and natural bridges, especially at the Navajo-Kayenta interface. Three examples of outstanding Navajo Sandstone spans are Rainbow Bridge, Pritchett Arch in Behind-the-Rocks country, and Double Arch in Arches National Park. One example of a Navajo spring-seep cave that eventually became a natural span is Bowtie Arch, on the Corona-Bowtie hiking trail near Moab.

Unconformity. An interval of erosion removed some late Triassic deposits from some eastern areas of canyon country.

Jurassic Period.

Unconformity. An interval of erosion removed early Jurassic deposits from canyon country, particularly in the eastern part.

Carmel Formation: siltstone, sandstone, shale and limestone, of shallow marine and tidal flat origin. This rock reaches a thickness of 650 feet to the west and north of canyon country, but where exposed in the eastern and central part of the region is generally a thinner layer that stands out in sharp color and texture contrast on top of the harder, white Navajo Sandstone. It was deposited along the coastal areas of a shallow, retreating sea, with its eastern edge about where the present Green River flows. To the east of the Green River, the Dewey Bridge member of the Entrada Sandstone lies on top of Navajo Sandstone. To the east the San Rafael Swell, the Carmel Formation is reddish in color, but is gray-green or yellow-gray to the west because of changing mineral content.

Since the Carmel Formation is composed of alternate harder and softer layers, it erodes similar to the Moenkopi and Cutler formations. At its edges, the Carmel often stands on a shelving base of hard white Navajo Sandstone, while supporting vertical buttes, walls or columns or salmon-hued Entrada Sandstone. Examples of this are visible almost anywhere that Entrada Sandstone still exists above the Navajo.

Carmel unprotected by the harder, overlying Entrada is not as resistant to erosion, but still occurs in many places, forming vast, colorful deserts. One such is to the east of Halls Crossing Marina on

Lake Powell, surrounding the approach road, Utah 263. Western and southwestern areas of the San Rafael Swell have vast expanses of Carmel desert, and the same formation tops the Navajo in the Escalante area, reaching eastward from the base of the Kaiparowits Plateau. There are also numerous large exposures of Carmel red desert to the west of Canyonlands National Park.

Entrada Sandstone. Entrada Sandstone consists of three distinct tongues or members, although only one of these extends into western canyon country.

Entrada/Dewey Bridge member: red siltstone, of marginal marine mud-flat origin, with deposits up to 100 feet thick. The sediments that formed this relatively thin rock layer came from the east, to interlayer with the Carmel Formation in the vicinity of the present Green River and on south. Dewey Bridge rock looks and erodes like the eastern-most Carmel deposits. Until recently, they were considered to be the same formation. The distinction between them is largely of interest only to geologists.

Exposures of the Dewey Bridge member occur near the paved road that traverses Arches National Park, and at the base of the Monitor and Merrimac buttes to the north of Utah 313, after that highway climbs up out of Sevenmile Canyon. Some exposures of Dewey Bridge are severely contorted, such as the examples just noted. This phenomenon is discussed in a later chapter under "wrinkled rock."

Entrada/Slickrock member: sandstone and siltstone, of desert-dune and marine tidal-flat origin. This member of the Entrada is dominantly reddish-colored desert-dune sandstone in the eastern and southern part of canyon country. It phases into less colorful siltstone in the northwest part of the region, and varies up to 350 feet in thickness. In the southeast corner of canyon country, it loses its color and is almost pure white.

In local areas, coastal lagoons have added different colored horizontal layers to the salmon-colored desert-dune rock. This same water-layering appears at the top of the Slickrock member. In the northwestern area of canyon country, the coastal siltstone Slickrock member closely resembles the underlying Carmel Formation.

The colorful desert-dune part of this Entrada member erodes into convoluted cliffs and, on its upper surface, into rounded fins, spires and domes. Red bluffs and fins of this picturesque sandstone dominate eastern canyon country from a few miles south of Interstate 70 almost to Monticello, and from the Green-Colorado river vicinity eastward into Colorado. Within this area, this sandstone erodes easily into arches, especially at the Slickrock-Dewey Bridge interface. Many of the spectacular natural spans in Arches National Park are in the Entrada Slickrock member. So is Wilson Arch beside U.S. 163, south of

Dewey Bridge member, between a Navajo Sandstone base and Entrada Slickrock cliffs.

Entrada Slickrock member, Rainbow Rocks.

La Sal Junction.

There are many other outstanding exposures of the desert-dune part of the Entrada Slickrock member, such as the Fiery Furnace, Klondike Bluffs, and Devils Garden areas in Arches National Park; the red bluffs near U.S. 163 south of Moab/Spanish Valley and on to the U.S. 163-Utah 211 junction; the massive sheer-walled buttes to the north of Entrada Bluffs Road, as it continues eastward from the Dewey Bridge; and, the upper sheer cliffs in Colorado National Monument, near Grand Junction, Colorado.

The siltstone part of the Slickrock member is exposed in many places in the general vicinity of the San Rafael Swell, but the best place to view its unusual erosional characteristics is in Goblin Valley State Park, where weathering has created a valley full of picturesque "goblins," or oddly-shaped balanced rocks and columns.

Entrada/Moab tongue: white dune-sand, of sea-coastal origin, up to 150 feet thick. This tongue of the Entrada occurs only in the northeast quarter of canyon country. It layers into the Summerville Formation to the west and south, and was deposited by an area of coastal dune sand.

The Moab tongue or member of the Entrada erodes on its upper

52

Moab tongue, Klondike Bluffs.

surface into the low profile "petrified dunes" typical of wind-deposited sandstone. Along edges, it breaks away into near-vertical low cliffs, which may be rounded by weathering. Where it is exposed extensively near a salt valley upthrust, such as Salt Valley in Arches National Park, extensive cracking caused by the upthrust has led to the formation of countless sheer-walled domes and fins of Entrada Slickrock and Moab tongue sandstones, with the younger ones still capped by the white Moab tongue sandstone, and with still others slowly forming in the white Moab tongue slopes to the east and west of the valley.

The Moab tongue is also visible as the white band on top of the salmon-hued Entrada Slickrock member in the upper cliffs of Wilson and South mesas on the lower slopes of the La Sal Mountains to the southeast of Moab, and in Colorado National Monument and on south along the Colorado-Utah border. The white-topped fins and slopes of the Klondike Bluffs can also be seen to the east of U.S. 163, in the vicinity of the Canyonlands air field. The white rock is Moab tongue sandstone.

Curtis Formation: light gray sandstone and greenish-gray shale, of marine origin. These sediments were deposited by a shallow sea invading canyon country from the north, but only into its northwest corner, in the general vicinity of the San Rafael Swell. The Green-Colorado rivers and Henry Mountains mark the eastern and southern extents of this formation. It varies up to 250 feet in thickness and layers into the lower Summerville Formation to the south and east in canyon country, along the margin noted.

Within the San Rafael Swell vicinity, the Curtis is just above the Entrada Slickrock member, but a short period of erosion separated the two intervals of deposit. Some of the sediments that make up the Curtis Formation came from the newly uplifted region to the west of canyon country, in central Utah.

The Curtis Formation erodes into ledges and weather-rounded shapes. It is exposed many places in the vicinity of the San Rafael Swell. Interstate 70 goes through a band of Curtis in the western reaches of this uplift, but the best place to view it is in Goblin Valley State Park, where it forms the upper strata of the valley.

Summerville Formation: interlayered sandstone, siltstone, mudstone, shale and gypsum, of a coastal-marine mudflat and tidal basin origin. This formation is predominantly red or red-brown in color, with some light tan or greenish layers, and is up to 330 feet thick. The lower part of the formation interlayers into the intruding Curtis Formation and Entrada Moab tongue, but the upper formation occurs throughout canyon country. It formed in the tidal basins and mudflats of the retreating Curtis sea.

On its edges, the Summerville usually erodes into ledgy cliffs because of its layers of varying hardness, but on its upper surface it weathers into colorful, open desert, with exposures of ripple rock and mud-crack casts.

Because open exposures of Summerville sediments erode away fairly rapidly, most Summerville in canyon country is found in cliff faces. The lower parts of the convoluted bluffs to the north of the town of Bluff are Summerville. Similar cliffs can be seen many places in the outer perimeter of the San Rafael Swell.

More open exposures of Summerville are scarce in canyon country, but I-70 goes through a half-mile wide band just west of the Curtis Formation exposure in the western part of the San Rafael Swell, and there are broad exposures to the south and west of Goblin Valley that can only be reached by off-road vehicle or hiking. In Arches National Park, the narrow bands of red sediments that top the white Moab tongue slickrock on each side of Salt Valley are Summerville.

The Summerville Formation is the last and youngest of the great red bed sediments that were deposited in canyon country at intervals over many millions of years, beginning in the Pennsylvanian period.

Bluff Sandstone: light-colored sandstone, of desert-dune origin, that is 250 feet thick at its maximum. This sandstone was deposited by desert sand dunes that followed the retreating Curtis sea northward, but only to about the latitude of Blanding. North of this line, it interlayers into the Summerville Formation, after covering some of these coastal mudflats. Thus, Bluff Sandstone occurs only in the southeastern part of canyon country. There are closely similar sandstones called Junction Creek, Winsor and Cow Springs in southwestern Colorado, south-central Utah and north-central Arizona, respectively.

Bluff Sandstone erodes into low, light-colored bluffs or terraces of weather-rounded rock. It is exposed to easy viewing in the vicinity of Bluff, where it forms the upper parts of the convoluted bluffs to the north of town and on upriver. The uppermost hard sandstone layer through which U.S. 163 descends from the north to reach Bluff is also Bluff Sandstone, and the lower several miles of Recapture Creek, to the east of Bluff, flows through the same rock. In the vicinity of the Recapture Creek-San Juan River confluence, there are extensive exposures of the upper surface of this light-hued sandstone.

Bluff Sandstone, where it exists, and elsewhere the Summerville Formation, are the uppermost units of the San Rafael Group, which includes all Jurassic strata from the Carmel and Dewey Bridge up through the Bluff and Summerville. Bluff Sandstone is also the uppermost of the several major desert-dune sandstones that covered all or part of canyon country. It marks a drastic change in climate and terrain in the region.

Morrison Formation. The Morrison Formation consists of four members, two of which barely project into the southeast corner of canyon country. The two main members together covered the eastern half of Utah, including virtually all of canyon country, plus all of Colorado and Wyoming, major parts of Montana, Nebraska, New Mexico and the Dakotas, and bits of Arizona, Texas, Oklahoma and Kansas.

The immense freshwater lake and stream region that deposited the Morrison Formation was ideal habitat for the many dinosaur species that dominated the land at that time. Fossilized dinosaur bones are fairly common in Morrison deposits. There are outstanding examples of petrified bone accumulations at Dinosaur National Monument and at the Cleveland Lloyd Dinosaur Quarry south of Price. There are other such locations presently being excavated or known.

The entire Morrison Formation contains volcanic ash, which adds a gray hue to its various sedimentary strata. The Morrison Formation contains uranium, but the lower Salt Wash member contains more than the upper Brushy Basin member.

Morrison Formation, I-70/U24 junction.

Morrison/Salt Wash member: light colored sandstone interbedded with red and green shales, all of lake, stream and floodplain origin, with a liberal admixture of volcanic ash. This member makes up the lower half of the Morrison Formation and is up to 350 feet thick. Salt Wash deposits are essentially an immense, broad alluvial fan of freshwater sediments from the southwest in Arizona, with some material from higher areas to the west in central Utah.

The Salt Wash member erodes into alternating harder ledges of light colored sandstone, and slopes of softer, colorful shale. The upper part of this member is a thick, light-hued layer of sandstone.

In the Four Corners vicinity, another smaller member of the Morrison, the Recapture, replaces the upper part of the Salt Wash. The Recapture member contains sandstone, conglomerate and shale, and erodes into more uniform slopes than the Salt Wash member. Recapture sediments came from the south.

Exposures of Morrison/Salt Wash sediments occur many places in canyon country, but most are not readily accessible except to off-road vehicles. One exposure convenient to viewing is at the Interstate 70-Utah 24 junction and on south beside U 24 for about three miles. In the Yellowcat area, east of Arches National Park, the lower terrain adjacent to the Yellowcat trail is Salt Wash sediments. In Montezuma Canyon, south of Monticello, the ledgy slopes just above the slickrock canyon walls are Salt Wash, as are the northeastern slopes of the northern reaches of the Uncompahgre Plateau in Colorado.

Morrison/Brushy Basin member: banded red, green, gray and purple shales, with some interlayered sandstone and conglomerate, all of lake, stream and floodplain origin, with a heavy admixture of volcanic ash from the northwest. Deposits occur up to 450 feet thick. This member makes up the upper half of the Morrison Formation. Brushy Basin sediments also came from higher land to the southwest and west of canyon country.

The Brushy Basin member erodes into color-banded hills and slopes. Larger exposures are called "painted desert." This member is characterized by colorful slopes between the hard upper sandstone layer of the underlying Salt Wash member, and the harder sandstones of either the Burro Canyon Formation in eastern canyon country, or the Cedar Mountain Formation to the west.

In the Four Corners vicinity, the lower part of the Brushy Basin is replaced by the Westwater Canyon Sandstone member of the Morrison. This member contains red and gray sandstone, shale and conglomerate from farther south, and in New Mexico is heavily mineralized with uranium.

Brushy Basin shales contain another mineral carried into the region in volcanic ash, a mineral that has the curious property of expanding and contracting with wet-dry cycles. This leaves exposed surfaces "frothy" in texture.

There are scattered exposures of Morrison Brushy Basin many

places in canyon country, but most are accessible only to off-road vehicle. One exposure convenient to viewing is east of U.S. 163, just north of the U.S. 163-Utah 313 junction, where typical Brushy Basin painted desert hills stand above open desert. The banded slopes above the Yellowcat trail, in the Yellowcat area east of Arches National Park, are also Brushy Basin, as are the higher slopes above Montezuma Canyon south of Monticello. To the west of this picturesque, historic canyon, Recapture Creek flows through this member between U.S. 163 and Utah 262. This canyon is accessible to hikers.

Cretaceous Period.

Burro Canyon and Cedar Mountain Formations: sandstone, shale and conglomerate, with some mudstone in the Cedar Mountain and more shale and limestone in the Burro Canyon, but all of lake, stream and floodplain origin. These two formations are essentially the same, with deposits to the east of the Colorado River being called "Burro Canyon," and to the west "Cedar Mountain." The shale and traces of limestone in the eastern area indicates more shallow brackish lakes there than to the west. The sandstones are generally light brown, while the shales are shades of gray, purple and green. Maximum thicknesses vary from 260 feet for Burro Canyon, to 500 feet for Cedar Mountain.

The sandstone layers of each resist erosion and form the hard ledges just above colorful Morrison/Brushy Basin slopes. Burro Canyon shales erode into the smoother slopes typical of most shales, and contain a mineral that swells with moisture, as in the Morrison/Brushy Basin. In the San Rafael Swell area, the Cedar Mountain Formation has a thin, lower member called Buckhorn Conglomerate which is coarse and gravelly in appearance.

These two early Cretaceous formations are not widely exposed in canyon country because most Cretaceous and late Jurassic rock has been removed from the region by massive erosion within the last ten million years. In general, one or the other of these two formations can be seen just above Morrison deposits wherever these are exposed. One of the most accessible exposures of the Burro Canyon Formation is in Lisbon Valley, southeast of La Sal, where it dominates the eastern slopes, as well as the higher slopes above nearby East Coyote Wash.

A small area of the Cedar Mountain Formation is visible to the east of U.S. 163 not far north of the U.S. 163-Utah 313 junction, on top of the painted-desert Morrison slopes and on to the north. Interstate 70 goes through a two-mile exposure of Cedar Mountain just west of the Morrison exposure in the western part of the San Rafael Swell. The largest exposures of this formation in canyon country are in the

58

northern part of the San Rafael Swell, in the general vicinity of the Cleveland Lloyd Dinosaur Quarry and Cedar Mountain, after which this formation was named. Other exposures are visible to the west of U.S. 6 on the San Rafael Swell slopes, beginning about 10 miles north of the I-70/U.S. 6 junction.

Unconformity. An interval of erosion removed some of the last Burro Canyon and Cedar Mountain deposits, although in some locations these formations blend inconspicuously into the Dakota Formation just above.

Dakota Formation: sandstone, with some interlayered shale and carbonized organic matter, of advancing seacoast origin. The sandstone is light brown to whitish, the shale gray. The carbonized layers that occur in some places are black. The whole formation attains a maximum thickness of 350 feet.

The Dakota Formation consists of three more or less distinct layers, all formed by a sea advancing into canyon country from the southeast. The first is primarily the erosion products and upper sediments of the Burro Canyon and Cedar Mountain formations just below, mixed up along the shores of the advancing sea. The next layer that phases in is shale from tidal lagoons and marshes. In some areas, vegetation and other organic material in such lagoons eventually converted to coal or carbonized matter that is not quite coal. The uppermost layer of Dakota is sandstone composed of beach sand from the advancing sea. All three layers may not occur in any one place.

Dakota sandstone erodes into low cliffs or bluffs, with the shale forming slopes where it occurs. In some areas, exposed terraces of the harder upper surface of this formation dominate the terrain. Massive erosion within the last 10 million years has left little Dakota and more recent sedimentary rock in canyon country, but there are some exposures that can be viewed.

In the Yellowcat region east of Arches National Park, the long east-west ridge has Dakota on its northern slope, just south of the grayish, rolling hills of Mancos Shale. This ridge continues east to cross the Colorado River north of the Dewey Bridge, and Dakota sandstone occurs there too, within sight of Utah 128. The lower slopes of East Coyote Wash, southeast of La Sal, are Dakota, and the hard, white upper layer of this formation occurs on both sides of U.S. 163 at the top of the winding grade that begins about 6 miles north of Monticello.

To the north of Moab, a narrow strip of light-colored Dakota lies to the east of U.S. 163, just beyond the open hilly desert, in the general vicinity of Canyonlands air field.

Interstate 70, in the western reaches of the San Rafael Swell, goes through bands of formations, as noted in the earlier pages. Dakota sandstone lies just west of the Cedar Mountain and Morrison formations.

59

Mancos Shale: gray shale, with thin layers of sandstone and limestone, of marine origin, up to 3500 feet thick. This distinctive shale was deposited by a stagnant sea that slowly advanced into canyon country from the southeast, fluctuated in level for millions of years, then slowly retreated.

Mancos deposits are principally gray shale that weathers into smooth slopes or rounded mounds, but as the sea that deposited this shale slowly advanced, rose and fell, then fitfully retreated, sediments from land to the west interlayered with the marine deposits, forming several distinctive eastward tongues. These bear such names as Tununk Shale, Ferron Sandstone, Blue Gate Shale, Emery Sandstone and Masuk Shale, but most of these intrude only into the western part of canyon country.

On the Kaiparowits Plateau in the southwest corner of canyon country, Tropic Shale is equivalent to Mancos Shale. Tropic Shale is topped by Straight Cliffs Sandstone. Both were deposited in what was once immense Kaiparowits Basin.

The various members of Mancos Shale are exposed along the base of the Book Cliffs to the north of Interstate 70 and to the east of U.S. 6, between the Henry Mountains and Waterpocket Fold, and in the western reaches of the San Rafael Swell. There, Utah 10 travels through long stretches of Blue Gate Shale between Price and the I-70/ U 10 junction. There are wide bands of Ferron Sandstone and Tununk Shale to the east of U 10. I-70 cuts through all five of the Mancos Shale members as it leaves the San Rafael Swell in the I-70/U 10 junction vicinity.

Undivided Mancos shale is exposed over broad areas of northern canyon country, as the gray, mounded desert that surrounds much of Interstate 70 between Grand Junction and Green River, then adjacent to U.S. 6 on north to Price. It also occurs in patches in the general vicinity of Monticello, and in isolated valleys in southwestern Colorado. The northernmost few miles of U.S. 163 also pass through a Mancos Shale exposure.

Tropic Shale and Straight Cliffs Sandstone are visible in the eastern cliffs of the Kaiparowits Plateau, from a back country road that goes southeast from Utah 12, between Boulder and Escalante.

Mesa Verde Group. This term is loosely applied to a broad range of formations, tongues and members of strata adjacent to canyon country, but entering this region only on its fringes. The group includes such strata names as the following. To the southeast, in the vicinity of Mesa Verde National Park: Point Lookout Sandstone, Menefee Formation and Cliffhouse Sandstone. To the north and northeast, in the vicinity of the Book Cliffs: various members of the Price River Formation, Farrer Formation and Neslen Formation. To the northwest and west, in the Price vicinity and south: Price River Formation, Castlegate Sandstone and others. To the southwest, Straight Cliffs Sandstone, and the Kaiparowits Formation farther west, are rough equivalents to Mesa Verde Group strata.

All of these many distinct rock layers have one general factor in common, hence their designation as a group. They were all deposited as coastal sediments as the Cretaceous sea slowly and fitfully retreated from the region. Some are shallow, near-shore marine sediments, some are beach sand, some are lagoon and swamp sediments heavy with organic matter turned wholly or partially to coal, and others are fresh water conglomerate deposits made on or near the seashore.

Many of the Mesa Verde Group rock strata are intricately interlayered with Mancos Shale sediments along the sea's western shoreline region, now the westernmost fringes of canyon country, and with each other. In general, Mesa Verde strata are older toward the west and younger toward the east, as the sea retreated across eastern Utah, pursued by freshwater sediments from the highlands of central and western Utah.

Eventually, Mesa Verde freshwater sediments of one kind or the other covered the Mancos Shale marine sediments throughout canyon country and on to the east and south into Colorado, New Mexico and most of Arizona. During the late Tertiary Period, Mesa Verde Group and equivalent deposits were removed from most of the canyon country region by erosion, leaving them principally in the Book Cliffs, plus the Straight Cliffs Sandstone on the Kaiparowits Plateau.

Various strata of Mesa Verde rock are visible as alternating layers of harder sandstones and softer sediments in the lower Book Cliffs to the north of Interstate 70, between Grand Junction and Green River, and to the east of U.S. 6, between I-70 and Price.

Unconformity. An interval of erosion removed some upper Mesa Verde deposits in some areas, but not all.

Wasatch Group. This term is applied to three major rock strata: the North Horn Formation, Flagstaff Limestone and the Colton Formation, plus other related layers to the east. These strata consist of mudstone, siltstone, sandstone, shale and limestone, all of freshwater sedimentary or brackish lake origin. All were deposited during the late Cretaceous and early Tertiary periods by great freshwater lakes to the north and west of canyon country. The northern lake filled the Uinta Basin, which formed during the late Cretaceous, and the western lake accumulated the colorful deposits now visible in Bryce Canyon National Park and Cedar Breaks National Monument.

Since subsequent erosion removed these deposits from most of the canyon country region, they now occur only within the Book Cliffs area. Some can be viewed distantly from Interstate 70 and U.S. 6, but for closer examination it is necessary to penetrate the Book Cliffs maze via various roads and off-road vehicle trails.

Tertiary Period.

Wasatch Group. This group, as discussed in the preceding paragraphs, extended into the Tertiary period.

Green River Shale: shale, of freshwater lake origin, again deposited by a huge lake in the Uinta Basin to the north of canyon country. Southerly deposits of Green River Shale entered canyon country, but have been removed by more recent erosion except in the northern fringe of the region. There, Green River Shale makes up the uppermost reaches of the Book Cliffs area, but can only be viewed by entering the area via various roads and off-road vehicle trails. One central layer of Green River Shale is oil shale, and other types of oil occur in Green River Shale farther north, in the present Uinta Basin.

There are no more Tertiary Period sedimentary deposits presently within canyon country, but a little late Tertiary lava entered the western edge of the region. Such igneous deposits can be viewed in Capitol Reef National Park on the higher ground adjacent to Utah 24, and in the general vicinity of Boulder and north along the Boulder Mountain road.

Igneous intrusives also pushed into older strata during the Tertiary. Some of these are now exposed by erosion. This subject is discussed more fully in a later chapter, under the heading "laccoliths."

Quaternary Period.

The only deposits in canyon country from the Quaternary Period are glacial moraines and localized wind and water sediments that are not yet consolidated into rock.

There were glaciers in all three canyon country mountain ranges, the La Sals, Henrys and Abajos, and also in nearby high country above 10,000 feet elevation, during the ice age that covered various parts of the North American continent for most of the Quaternary Period. Another cold interval that ended only last century brought glacier activity again to the La Sal Mountains.

Glacial deposits have long since eroded away in the Henry and Abajo mountains in the 10,000 years since they were laid down, but glacial moraine is still plentiful in the La Sals. Several off-road vehicle

Quaternary dune sand, White Wash.

trails traverse moraine areas, such as the Dark Canyon and La Sal Pass trails. In western canyon country, the dirt road that travels the flanks of Boulder Mountain between Torrey and Boulder goes through patches of glacial moraine.

Quaternary water deposits take the form of compacted sediments in canyons and valleys. In many places, such recent sedimentary layers have been cut by changes in erosion rates brought on by human activities. This subject is discussed further in a later chapter on major events that shaped canyon country.

Quaternary airborne deposits take the form of sand, dust and soil layers, some of them held more or less in place by plant life of various sorts. One microscopic community of plant life does such a good job of soil-holding in canyon country that sand dunes free to move about are relatively rare. Such "cryptogamic" soil is discussed further in a later chapter. An outstandingly beautiful area of "living" sand dunes can be seen in White Wash, near the Green River. The White Wash off-road vehicle trail enters the dunes area. There are many other dune areas in canyon country, such as where Utah 24 goes through the San Rafael Desert between Interstate 70 and Hanksville.

MAJOR SHAPING EVENTS

GENERAL

Previous chapters have discussed the general flow of geologic history in the canyon country region and the various sedimentary rock strata that were deposited there during the last half billion years. This chapter will discuss the major geologic events that shaped the sedimentary strata into the canyon country of today.

It should be emphasized, before discussing major geologic events and processes of any kind, that on the scale of human lives, or even the lives of whole human cultures, geologic events and processes are generally very, very slow. Most are too slow to observe by human senses within a single human life span. Some are too slow to be apparent within the full time span of human civilization.

Geologic and biologic processes are on entirely different time scales. Individual life forms may live from a few days to several decades. A species may exist relatively unchanged for hundreds of thousands, even millions of years. But a single geologic event, such as the thrusting up of a mountain range, may take 100 million years, and be so slow that erosion wears down the mountains as fast as they are thrust upward. A sea may take 10 million years to invade a region of continental land, stay there for 45 million years, then take another 10 million years to retreat from that land.

Thus, the average human observer sees an apparently static, unchanging world throughout a lifetime, and can easily conclude that what he sees now, always was and always will be. This is because a human life span is simply too short an observational base from which to detect most geologic processes.

Similarly, to certain insects that live for only a few days during the spring or summer months, the whole universe is one of perpetual sunshine and flowers, with an entire lifetime being too short for observing slower events such as seasonal cycles, or the relatively sedate pace of human affairs.

There are, of course, a few major geologic events and processes that are fast enough for ordinary human observation. Earthquakes and volcanic events are often blindingly fast, sometimes literally explosive, even on the human time scale. Even so, crustal slippages along earthquake fault lines, or the building of long chains of volcanoes, are geologic processes that may take 100 million years or more.

Surface erosion, by wind and water, is also a process that is fast enough to see in some areas, within the span of a human life. Canyon country is one such area.

Because most geologic events and processes function on a different and vastly slower time scale, they are generally detectable only by careful, long-range scientific observation, or by inference from slowly accumulated scientific data. Geologists can read the stories told by sedimentary strata, but they also use many other scientific methods for detecting, measuring and understanding the slower geologic events. These methods are discussed in detail in some of the books listed under Further Reading.

In order to gain a general understanding of the geologic events and processes that shaped canyon country as it is today, it is necessary to go back in time at least 500 million years. For a full understanding, it would be necessary to go back even further, perhaps for another three billion years, but the scope of this book is not that broad.

For even a general understanding, however, a basic knowledge of current continental drift theory is necessary, because geologists are discovering that this new concept can explain many of the geologic events observed throughout geologic history, events that had no known cause before the recent general acceptance and further refinement of the continental drift concept. A brief summary of this concept appears in the following pages.

Throughout the geologic history of canyon country, various continental drift events have occurred. These have triggered other geologic events in the western part of the North American continent. Some of these have, in turn affected canyon country, either directly or indirectly. Thus, the full sequence of geologic cause-and-effect that led to present-day canyon country must begin with continental drift events, even though the study of such relationships is still very new.

Of course, other more subtle factors have also affected the shaping of canyon country, such as solar flares, sunspot activity and other astronomic phenomena, but for the most part such factors act indirectly, by affecting the Earth's climate or weather. Thus, while a major rise in sea level over several million years may primarily be caused by a continental drift event, other more subtle factors, such as a long period of increased solar activity or the development of major species of shallow marine life, may also contribute to the overall geologic effects produced by the higher sea level.

Although the scope of this book limits discussion to the major shaping events, it should be understood that in few cases is the sequence of geologic cause-and-effect simple and uncomplicated. However, most such complications are of interest only to geologic scientists and in no way detract from the broad, general picture of canyon country geology presented in this book. The various books listed under Further Reading, plus many others, provide a wealth of information for those who wish to delve more deeply into the fascinating subject of canyon country geology.

CONTINENTAL DRIFT THEORY

The following brief summary of continental drift theory is, of necessity, quite limited, but is complete enough to provide a general understanding of the major shaping events described later. Popular literature on the continental drift concept is scarce, because the concept has been widely accepted by geologists only within the last ten years. Some of the books listed under Further Reading contain excellent descriptions of the basic idea.

PLATES. The Earth's crust is divided into a number of caps or "plates" of different sizes and shapes. There are two types of planetary crust, oceanic floor and continental. Continental crust is almost twice as thick as oceanic crust and is lighter, or less dense. Plates may be all oceanic crust, or partly oceanic and partly continental, but not all continental. Plates vary in size. The Pacific plate presently covers about one-fifth of the Earth's surface. Others are smaller. A few plates cover only a few degrees of the surface. Oceanic crust is temporary. It is created and lost at the edges of the plates. Once created, continental crust is permanent.

PLATE MOVEMENT. The various plates on the Earth's surface move constantly, relative to each other and the Earth's poles, in different directions and at different rates of speed. The directions and rates of movement change at times. The reason why the plates move is unknown at present, but geologists agree that the immense energy needed to power this movement comes from heat within the Earth's interior. Continents, together with the oceanic crust parts of their plates, move at rates that vary from 1 to 8 inches per year.

CONVERGING PLATES. If two oceanic crusts are going toward each other, one goes under the other and is shoved steeply downward into the Earth's interior. If oceanic and continental crusts are going toward each other, the oceanic plate always goes under. When oceanic crust goes downward into the Earth's interior, this is called "subduction." See Figure 6.

If two continental crusts come together, neither goes down, but the impact creates an immense mountain range. Geologically recent examples of this are the Himalayas and Alps. The Himalayas are still growing. Older examples are the Appalachians and Urals.

The permanent continental parts of plates drift slowly about on the Earth's surface, coming together in various combinations, then later breaking apart into differently shaped continents. The present continents are only the most recent combination. New continental crust is created in bits and pieces by involved processes. Smaller separate pieces are later shoved together with larger land masses by the oceanic plate subduction process.

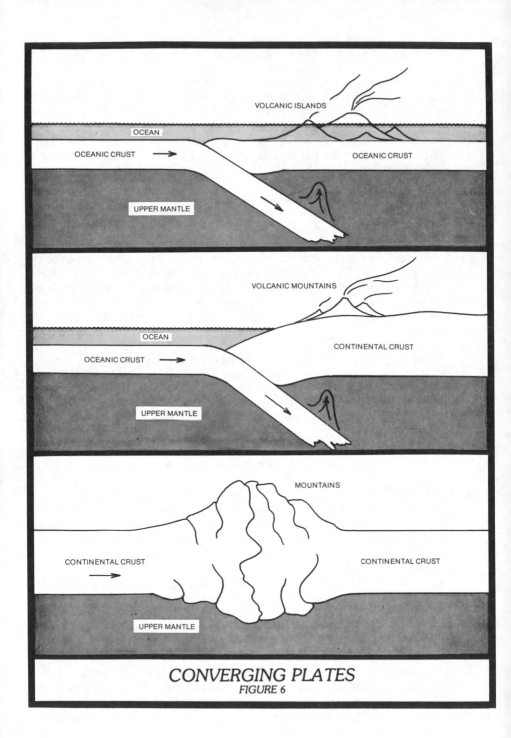

CONVERGING PLATES
FIGURE 6

68

DIVERGING PLATES. Most diverging plate junctions involve oceanic crust and are on ocean floors. At such junctions, the oceanic crusts on each side of the roughly straight junction separate. The gap is filled by molten rock from the Earth's interior, thus creating new ocean crust to make up for that lost by subduction elsewhere. The process of making new ocean crust is called "sea floor spreading." Typical rates of spreading vary up to 8 inches per year and change from time to time. Because oceanic crust is continually being formed by sea floor spreading, and destroyed by subduction elsewhere, such crust is temporary. The oldest oceanic crust of Earth is only about 200 million years old, while the oldest continental rock is billions of years old.

New diverging plate junctions that occur in continental crust first form "rift valleys" then, as they open wider, become narrow seas and ultimately oceans. The great rift valleys of eastern Africa, the Red Sea and the Atlantic ocean are current examples of these three stages of continental crust separation.

PASSING PLATES. In some places, adjacent edges of plates neither converge nor diverge, but slide past each other. Such sliding junctions are called "transform faults." The San Andreas Fault in California is a highly visible example of such a fault. It runs from the Gulf of California to San Francisco. The land to the west of the fault line, including all of Baja California and the California coast up to San Francisco, is on the Pacific Plate. The land to the east of the fault is on the North American plate. Because of the relative directions of movement of these two plates, 50 million years from now Baja California and a large segment of California west coast will form an island chain in the north Pacific, due south of Alaska.

ACTIVITIES AT PLATE EDGES. Various major geologic activities take place at plate edges, again depending on their compositions and directions of relative movement. Above subduction zones, volcanoes build mountain ranges on land, or chains of volcanoes and volcanic islands in the ocean. The well-known "Ring of Fire" around the Pacific Ocean is comprised of volcanoes along lines of present or recent plate subduction. The Andes Mountains and Japanese Islands are examples of volcanic ranges and islands built over subduction zones.

Zones of sea floor spreading also build higher terrain, but as newly formed oceanic crust cools, it shrinks and lowers. A continuous chain of such oceanic ridges or mountains some 47,000 miles long lies on the floor of the Earth's oceans. There are also many other shorter oceanic ridges.

Earthquakes also occur where plates are forming by sea floor spreading, where they are subducting, or where their edges are slipping along transform faults. Sometimes subduction zones along continental eges, or new contacts between plates create lateral pressure on adjacent areas of the continent. These pressures can cause massive folding, faulting and uplifting, and even contribute toward inland volcanism and the formation of non-volcanic mountain ranges.

Subducting oceanic crust usually heads down into the Earth's interior at a fairly steep angle, eventually to remelt, but sometimes gigantic slabs of angled, subducting crust break off and stop going down. The continent then passes on over such detached slabs, and subduction zone effects such as volcanism appear far inland from the active line of subduction. Such a detached slab has contributed to the shaping of canyon country.

ACCELERATED SEA FLOOR SPREADING. Steady rates of sea floor spreading do not affect the Earth's sea level, but if the rate increases appreciably, the larger amount of new, hot oceanic crust creates new and large oceanic ridges or mountains. This in turn displaces ocean water and raises the Earth's sea level for the duration of the faster sea floor spreading rate. Once the rate slows again, and the new oceanic crust has cooled, the sea level lowers to approximately its former elevation.

Such changes in sea floor spreading rate have caused major changes in sea level within the Earth's geologic history. These in turn have caused some of the sea invasions of canyon country described in earlier chapters.

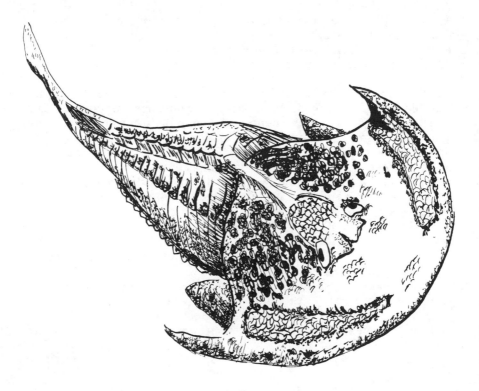

RECENT CONTINENTAL DRIFT HISTORY. Most of the continental drift events that have helped shape canyon country have occurred within the last 500 million years. Geologists are finding plenty of data for reconstructing this most recent chapter in Earth's continental drift history but for earlier chapters the clues are obscured by more recent events. Following are some of the highlight events within the last 500 million years. The dates given are approximate. The effects of some of these events on canyon country are discussed in other chapters.

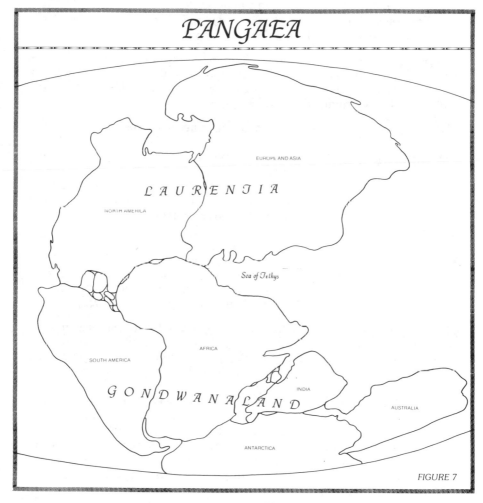

FIGURE 7

500 million years ago. The land mass that later became the North American continent joined the rest of the Earth's continental land mass to form a supercontinent that geologists have named "Pangaea." See Figure 7. Pangaea was not solid land, but was split into two major

71

masses that were partially in contact, but largely separated by a narrow wedge of water called the "Sea of Tethys." One mass, consisting of what is now North America, Europe and Asia, was named Laurentia. The other, consisting of what is now South America, Africa, Australia, India and Antarctica, was named Gondwanaland. As the North American continental land joined the others, the impact raised the Appalachian Mountains and their equivalents in the present British Isles, Norway and Greenland. At that time, and for the next 100 million years or so, Pangaea was centered on the Earth's south pole and was heavily icecapped in the polar region.

The rest of Earth was a vast, warm ocean. At that time, the equator ran through Utah on a northeast-southwest angle.

400 MY ago. The supercontinent Pangaea started drifting northward. As it did, its immense icecap slowly melted, and changes in sea level, ocean currents and temperatures affected the Earth's climate and biologic development in drastic ways.

250 MY ago. By this time the equator was angling across Texas.

225 to 180 MY ago. The Sea of Tethys widened, further separating Laurentia from Gondwanaland.

150 MY ago. The North American continent and some closely related continental land masses separated from Laurentia, leaving what is now Europe and Asia.

125 MY ago. The South American continent separated from Gondwanaland, leaving what is now Africa, India, Australia and Antarctica. These land masses later separated and went their various directions, with India heading north, eventually to impact with Asia.

81, 63 and 53 MY ago. At each of these times, the North American plate changed its direction and rate of separation relative to Europe and Africa. Each of these events also triggered secondary geologic events in the western part of North America, including canyon country.

38 to 20 MY ago. Prior to this interval, the western edge of North America had been overriding other oceanic plates, but from 38 to 20 million years ago, it made an unusual first contact with the Pacific plate. This contact formed the San Andreas transform fault system noted earlier, placed lateral crustal stress on the western part of the North American continent, including canyon country, and triggered secondary geologic events there.

20 MY ago to present. Since the San Andreas Fault formed, intermittent slippage along this fault has relieved pressures between the Pacific and North American plates and triggered still other secondary geologic events that have affected the western part of North America.

EVENTS, CAUSES AND EFFECTS

The following paragraphs note some possible relationships between continental drift events and geologic events and effects within western North America and canyon country. It should be understood that for the most part these suggested relationships are as yet purely hypothetical and not proven to the satisfaction of the majority of research geologists. Further research may prove some of these suggested relationships and disprove others.

It should also be noted that none of the suggested relationships between newly studied continental drift events and well-known continental geologic events has been invented by the author of this book. All have been proposed by various geologic researchers on the basis of existing, but still limited, research data. These proposed relationships have been included in this book, even though not yet proven, in order to provide readers some insight into the general causes behind the more easily observed events and effects that have shaped western North American and canyon country. Even though some of the noted relationships eventually prove to be incorrect, there is little doubt that various continental drift events are the basic causes of certain types of continental geologic events and the shaping of continental terrain. That principle is clear, even though many of the details are still obscure at this time.

The following geologic events, their possible causes and clearly visible effects are listed from the oldest to the most recent. The dates are in terms of millions of years ago, unless otherwise indicated. The dates are approximate and in most cases are compromises between the slightly different dates given by various geologic researchers for the same event. However, on the vastly slower geologic time scale, such slight difference are generally insignificant. What counts is that even the approximate dates established do tend to show startling relationships between continental drift events and geologic events on the interiors of continents, including canyon country.

Late Precambrian to about 100 MY. For this long time interval, more than 600 million years, canyon country was within the Torrid Zone, between the tropic of Cancer and the Equator. Its climate was thus generally quite warm, either tropical-humid or desert-hot, depending upon other factors. The types and amounts of sedimentary deposits and erosion rates in canyon country were affected by these warm climatic conditions. Approximately 100 million years ago, Utah and canyon country crossed the Tropic of Cancer and thus entered the temperate zone where it has been ever since.

From 500 MY to about 225 MY. During this long interval there was no separate North American continent. Most of the continental land mass that was later to become North America was joined to Pangaea, the single huge supercontinent that contained all of Earth's land mass. The various geologic events and processes that affected canyon country during this 275 million year interval were happening to Pangaea, not to a separate North American continent.

About 320 MY. At this time, the Ancestral Rocky Mountains started forming in Colorado and adjoining states to the north and south, and the Emery Uplift started rising in canyon country. These uplifts continued for millions of years, and strongly affected sedimentation in canyon country.

From 310 MY to 300 MY. The first low rise of the Uncompahgre Uplift on the eastern border of canyon country occurred during this interval. This and subsequent uplifts of the Uncompahgre were very important to canyon country because of the sediments contributed, and because they helped form the Paradox Basin and related geologic features.

About 295 MY ag. A second, much greater upthrust of the Uncompahgre began about this time, affecting canyon country as noted above.

From 225 MY to about 180 MY. During this interval, the Sea of Tethys slowly changed and widened, separating Pangaea into Laurentia and Gondwanaland. There was still no separate North American continent. During this interval, the following major events occurred. The plate movements that were enlarging the Sea of Tethys probably contributed somehow to these events.

At about 225 MY and 210 MY, the Uncompahgre Uplift was thrust further upward for the third and fourth times. From about 200 MY to 150 MY, the general canyon country vicinity was a vast Sahara-like desert, broken only by relatively short intervals of wetter climate. At about 190 MY, the present Sierra Nevada mountain range started rising in stages. This immense uplift separated canyon country from the great western ocean, and thus affected its climate, rates of sedimentation and erosion, and other geologic factors. From that time on, seas invading the present Utah-Nevada region came from either the north or south.

About 150 MY ago. The North American continent formed at this time by starting to separate from Laurentia, or what was thereafter to be known as Europe and Asia, or Eurasia. This westward thrusting of

North America immediately produced secondary but major geologic events in the continent's western region, primarily because of new eastward pressure against the continental crust. This pressure initiated an interval of crustal deformation and related volcanic activity that lasted for about 25 million years.

One result of the crustal deformation was a great ridge or chain of mountains that extended from the southern tip of Nevada on north to Alaska. During its early phases, this deformation was so severe that the continental crust in what is now western Utah and adjacent states was shortened from 40 to 60 miles by overthrust. This overthrust and related crustal uplift was named the Sevier thrust belt. The mountainous highlands thus formed further isolated canyon country from the regions farther west, restricted later seas invading from the north and south, and also contributed immense amounts of marine, coastal and freshwater sediments to canyon country.

Beginning about when North America began to thrust westward, its western region was affected by widespread volcanism, undoubtedly related to the new phase of continental drift. Because of its thicker crust, no volcanoes developed within the Colorado Plateau and canyon country, but sedimentary strata deposited there during this interval were heavily laced with airborne ash from volcanoes to the northwest.

About 125 MY ago. The South American continent formed at this time by separating from Gondwanaland, which became known as Africa after still other continental land masses had separated from it. The interval of volcanism in western North America that began when that continent headed west, ended about the same time that South America headed west. These two events may be related, or may be coincidental.

It is interesting to note that the present African continent will be further reduced in size by a huge land mass now starting to separate from its eastern side along the line of the great rift valleys there.

From 110 MY to 85 MY. Sea floor spreading along the Earth's great 47,000 mile long rift system drastically accelerated during this interval, greatly increasing the height and volume of the sea-mounts in the vicinity of the rifts. This gradually raised the sea level all over the world some 1500 feet, and inundated about 40% of the Earth's land. This flooding became the sea that invaded canyon country during the Cretaceous period. Other factors kept this sea level high for 20 millions years after the sea floor spreading slowed again about 85 million years ago. During this high sea level interval, the Sevier overthrust-uplift belt in western Utah and elsewhere was reactivated and its mountains were raised still higher.

Age of dinosaurs. It is not the purpose of this book to discuss prehistoric life forms, except as their remains and traces may become a part of the geologic scene, such as visible seams of coal or oil shale, or outcroppings of tar sands. However, since dinosaur tracks and bones

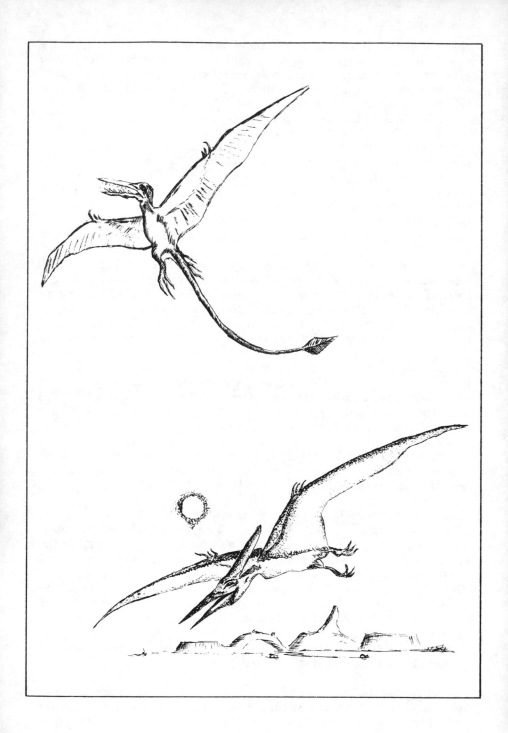

are a part of the canyon country geologic scene, it is appropriate here to mention that some scientists see a possible relationship between the above-noted interval of faster sea floor spreading and its flooding of continents and the resulting changes in global climate, with the ultimate extinction of dinosaurs, including the equivalent giant reptiles in the oceans.

Certainly, dinosaurs and their predecessors were around for a long time before the noted increase in rate of sea floor spreading, but the changed conditions brought on by the higher sea level were ideal for the development of more sophisticated forms of reptilian and amphibian life. Scientists conjecture that as the sea level fell, these ideal conditions disappeared, creating other conditions more ideal for the development of mammals, and unfavorable for large reptiles and amphibians. These highly specialized creatures were unable to adapt to the changing conditions and thus failed to survive. It should be noted that while the accelerated sea floor spreading ended about 85 million years ago, the seas took another 20 million years to recede to their former level. That point in time marks the final extinction of the dinosaurs.

The details of this interesting hypothesis are too complex for discussion in this book, but can be found in recent books and magazine articles on paleontology and continental drift.

About 81 MY ago. The North American plate changed direction and rate of travel at this time. Between 81 MY and 63 MY, this plate moved westward much faster than before or since. This accelerated plate movement and change of direction triggered a broad spectrum of geologic activity on and near the continent's westward, leading edge. Up to this time the Sierra Nevada range was essentially coastal. The accelerated plate movement buckled the off-shore oceanic crust to form the first upthrusting of the present California Coast Range, and added new lift to the Sierras.

Farther inland, the increased crustal pressure initiated an interval of violent geologic activity that lasted from 80 to 50 million years ago. Scientists call this interval of activity the "Laramide orogeny." Orogeny means mountain building and associated geologic activities. It may or may not include volcanic activity.

The Laramide orogeny brought tremendous change to western North America, including canyon country. The present Rocky Mountain range formed during this interval. Within canyon country and its borderlands, the Uncompahgre Uplift rose still higher, the Paradox Basin sank still lower, and the Monument, Circle Cliffs and San Rafael uplifts formed. These are discussed more fully in the following chapter.

About 63 MY ago. The North American plate changed direction and speed again, this time slowing. This change may have affected some of the Laramide geologic activities, but did not stop them.

About 53 MY ago. The North American plate changed direction again, but its rate of travel remained much the same. This change must have affected western North America plate conditions, because the Laramide orogeny ceased shortly thereafter, on the time scale of geologic events.

Henry Mountains.

About 48 MY ago. The Henry Mountains formed, when igneous material forced its way upward through faults in the lower crust of the continent and into the overlying sedimentary strata. Nearby Navajo Mountain may also have formed about this time. The fracturing of the thick Colorado Plateau crust that permitted these and other later igneous intrusions to form canyon country mountain ranges may have occurred during the Laramide orogeny, or may have occurred earlier in the Precambrian Era. The formation of the Henry, Abajo and La Sal mountain ranges and Navajo Mountain is discussed in more detail in the following chapter.

From 38 to 20 MY. About 38 million years ago, a change in plate relationships occurred on the west coast of North America. As the final bit of the smaller Farallon plate that had been subducting under the North American plate finally vanished, the immense Pacific plate made first contact with North America in the vicinity of what is now Mexico. This contact spread northward for hundreds of miles into California over the next 18 million years, and exerted enormous eastward pressure against the southwestern region of the North American continent.

This pressure produced a violent response. Faulting and volcanism started all over the whole western region, except within the thicker-crusted Colorado Plateau. There, two more intrusive mountain ranges formed, the Abajos and the La Sals, 28 and 24 million years ago, respectively. The Uncompahgre Uplift ws rejuvenated for the sixth time, and the San Rafael Swell was activated a second time, both uplifting still higher.

There is evidence to indicate that between 38 and 19 million years ago, coinciding with the interval of Pacific plate pressure, a detached slab of Farallon plate passed under the western states. Thus, the intense volcanism during this interval could have been caused by the detached slab of plate. Quite probably, both the pressure and the plate contributed toward the volcanism.

Whatever its cause, the volcanic activity ceased about 20 million years ago, at about the same time as pressure from the Pacific plate eased off.

From 19 to 16 MY ago. Volcanic activity in the region essentially stopped during this interval.

About 15 MY ago. About this time, the zone of contact between the Pacific and North American plates changed into a transform fault, the present San Andreas Fault, allowing the Pacific plate to slip northwestward in relation to the North American plate. This and subsequent slippages formed the present Baja California and the Gulf of California. The initial large-scale slippages also changed conditions at the plate interface by relieving all pressure. This in turn triggered a phase of crustal stretching in a vast region from the Wasatch Mountains in Utah to the Sierra Nevadas, and from northern Mexico to Oregon. Scientists are not in agreement as to what actually caused this stretching, but they do agree that it allowed great blocks of continental crust that had faulted earlier to tilt, forming a basin-and-range pattern, and to pull apart, allowing igneous rock to flow to the surface. This process has continued into the present, and has literally stretched the continental crust some 50 miles in this region.

Although the events that occurred during this interval of geologic activity had little or no effect upon canyon country, they no doubt would have if the crust of the Colorado Plateau had not been so much thicker than crust to the west. This interval is described in order to illustrate how the Colorado Plateau's thicker crust has protected it and canyon country from such violent geologic events as volcanism and

79

extensional block-faulting. In sum, throughout the last half-billion years of geologic history, it has taken extraordinary geologic force to deform and shape canyon country.

About 10 MY ago. At this time the region between the Sierra Nevada range and the Rockies and on to the east, began to uplift. Within a relatively short time, on the geologic time scale, the region gained approximately one mile of elevation. To date, geologists are at a loss to explain this immense uplifting of an entire region, mountain ranges and all, but from the amount of energy involved there is a good possibility that the uplift was caused by some aspect of continental drift or whatever sub-crustal forces motivate this drift. One hypothesis holds that the region was uplifted as it passed over a zone of previous sea-floor spreading. This concept is the most promising to date.

During this major uplift interval, the Uncompahgre Uplift was given its seventh boost, rising even more than the region surrounding it.

Water erosion, Arch Canyon.

The general regional rise initiated an entire new geologic phase in canyon country. Before the rise, the region had been lower, with gentle profiles. Even the nearby Rockies and other uplifts had much more rounded contours, and elevation differences were less pronounced than at present. The region's climate was temperate but fairly arid. Scant precipitation fell on the mountain highlands. No rivers or perennial streams crossed the open plains of canyon country.

The general uplift changed all that. The higher mountains and lowlands gathered more precipitation. The increased water flow formed rivers and streams that meandered across the gentle terrain. The present great canyon country river system was born, the Colorado, Green, San Juan, Dolores and their many tributaries. As the uplift continued, more water flowed and the meandering rivers and streams cut more and more deeply into their wandering courses, slowly but inevitably shaping the canyon country of today.

During the 10 million years since this great uplift started, canyon country rivers and their countless tributaries have removed about one mile of sedimentary rock from most of the region, slowly, impartially, uncovering the mountains, uplifts and lowlands, revealing them as they are today. This relatively rapid erosion process, begun 10 million years ago, is not yet complete, however. It continues even now, accelerated still further by a new factor — technological man.

About 1 MY ago. About this time, the ancient Uncompahgre Uplift, on the eastern flanks of canyon country, rose still again, for the eighth time.

From 800,000 to 10,000 years ago. During this interval, known as an "ice age," glaciers covered much of the North American continent, but did not reach as far south as canyon country. Within this region, however, separate glaciers did form on highlands above 10,000 feet elevation. These included the Abajo, Henry and La Sal mountain ranges and, on the western border of canyon country, the Aquarius Plateau.

During this major ice age, the increased precipitation and subsequent melting of glacial ice created Lake Bonneville in western Utah, a prehistoric lake that inundated about half of   western Utah to depths of 1,000 feet. Within canyon country, glaciers left their marks in the form of sharpened mountain peaks and ridges, deepened mountain valleys, large expanses of glacial moraine and more rapid cutting by the system of rivers and streams that drained the region.

Between 1200 and 1850 AD. Part of North America received more glaciation during this "little ice age." Within canyon country, only the La Sal mountains had glaciers during this very recent interval, although glaciers on the Aquarius Plateau to the west thrust into the region in the Boulder Mountain vicinity. This brief period of glaciation still further shaped the higher peaks and valleys of the La Sals, and also contributed to accelerated canyon cutting by the region's rivers and streams.

Early 1800s to the present. The entry of white men into the canyon country region initiated a new geologic event or process. Life forms have a long history of affecting geologic processes directly and indirectly. Examples of direct effects are the slowing of erosion by plant life, and by the forming of organic strata such as coal, oil, oil-shale, peat, tar-sands, lignite, etc., by various forms of life. Indirect effects are produced by slow changes in atmospheric composition and ground water chemistry, both of which affect the formation and metamorphosis of rocks. There are other even more subtle effects, but taken altogether, life forms have been a major geologic force over the ages, and will continue to be.

The human race is no exception to this. To the contrary, technology has turned humanity into a major geologic force all by itself. The effects produced by human activities are sometimes dramatic, but more often are too subtle to be apparent to untrained observers. In general, the human race affects geologic processes by affecting erosion rates, by moving or removing great quantities of materials, by affecting atmospheric composition and water chemistry, and by affecting other lifeforms so seriously that these, in turn, create their own range of effects.

Here are some examples of these human effects. Human construction and surface mining destroy ground cover plants and accelerate runoff and erosion. The removal of oil and water from underground reservoirs causes massive land subsidence, as does the underground mining of coal. All kinds of mining affect the chemistry of surface water and ground water, which in turn affect rock mineralization processes and life forms. The building of dams affects erosion rates, life forms, water chemistry, and weight distribution. This last can trigger land subsidence and even earth faulting and earthquakes.

Massive air pollution changes atmospheric chemistry, which in turn affects other life forms and water chemistry. Cloud-seeding and other weather-control efforts affect erosion rates, air and water chemistry and life forms. Long-range changes in atmospheric chemistry caused by the massive burning of fossil fuels may bring on another ice age, or a period of prolonged heat and drought. Farming and logging change erosion rates and affect atmospheric and water chemistry. The domestic grazing of arid land drastically accelerates erosion and affects other life forms. The creation and disposition of billions of tons of solid waste each year literally creates unique geologic strata, and also affects the air, water and life forms of the region in which it is deposited. The destruction of natural plant ground cover for any of hundreds of human purposes accelerated wind and water erosion and affects regional life forms.

In sum, most human activities, aided by technology, affect geologic processes either directly or indirectly. On the human scale of time,

Lake Powell, Glen Canyon National Recreation Area.

the effects seem relatively minor and slow, or observable only by scientists, but on the geologic time scale the human race has no counterpart for sheer speed of change. In effect, humans are an explosive geologic process, causing far more change within a specific interval of time than the most powerful of natural processes, the subduction of oceanic crust and the forming of mountain ranges.

Within canyon country, the human geologic process is having drastic effect, primarily because of the fragile nature of its arid-region life forms. Farming and cattle ranching have greatly accelerated erosion of the sandy soils by wind and water. This in turn has affected other life forms that depend upon the natural vegetation, and has initiated rapid stream and flashflood cutting into deep layers of geologically recent sedimentary deposits in valleys and canyons. Four readily accessible examples of this are Montezuma Canyon, south of Monticello; Salt Wash, upstream from Wolfe Cabin in Arches National Park; Cottonwood Canyon, west of Monticello, and upper Salt Creek Canyon, in the Needles District of Canyonlands National Park. There are countless others.

Mining, and bulldozer operations connected with mineral search activities, have severely damaged watersheds and created countless raw drainage lines for unhindered erosion. Great solar evaporation and mine tailings ponds have changed erosion patterns. Huge dumps of salts and other mineral wastes affect groundwater chemistry, and will eventually enter the river systems.

In settled areas, the human destruction of natural ground cover for various purposes has accelerated wind and water erosion. When it rains in such areas, rivers of topsoil wash away, and when winds blow, great clouds of dust rise where none rose before the cryptogamic soil was destroyed.

In the southern part of canyon country, human activities have introduced an entirely new element to natural erosion processes, that of large amounts of standing water in contact with sandstone. By its nature, sandstone is weakened by water, and drastically weakened by standing water, such as that in a reservoir. Thus, Lake Powell, for all its beauty, is dramatically accelerating erosion of its sandstone walls and shores, where dune sand is washed down into the depths and great masses of rock are sloughing off millions of years before they would have otherwise.

Unfortunately, the type of rapid erosion that standing water causes is also threatening such outstanding natural features as Rainbow Bridge. Another large arch has already been destroyed by this rapid, human-related erosion process.

There is ample reason to conclude, however, that the human race, being so destructive to the world upon which it depends for life, may not last very long as a geologic process. Many human activities, if continued for long into the future, could lead to the extinction of humanity as well as other life forms. Certainly, few knowledgeable scientists would predict that the human race will last nearly as long as the least of the dinosaurs did, making the human race a powerful but short-lived geologic process.

In sum, it would be safe to say that by one million years from now — a mere instant of geologic time — it will probably take a very skilled archeologist, indeed, to discover any remaining trace of the human race on this planet. By then, this old Earth will have shrugged off its transient tormentors and reverted to its own invisibly slow way of shaping its surface into endless varieties of natural beauty, perhaps for the enjoyment of some less destructive species.

MAJOR SURFACE
FEATURES

GENERAL

Earlier chapters have outlined canyon country geologic history and discussed its rock strata and the major events that have shaped them. This and the next chapter discuss the surface features that past geologic activities have produced and, in some cases, are still producing.

As each feature is discussed, if there are only a few such features in canyon country, all of these are listed. If there are too many of a particular feature for complete listing, representative examples are noted. As an aid to finding particular feature descriptions, they are listed in alphabetical order within each chapter, and again in a combination alphabetical index in the back of the book.

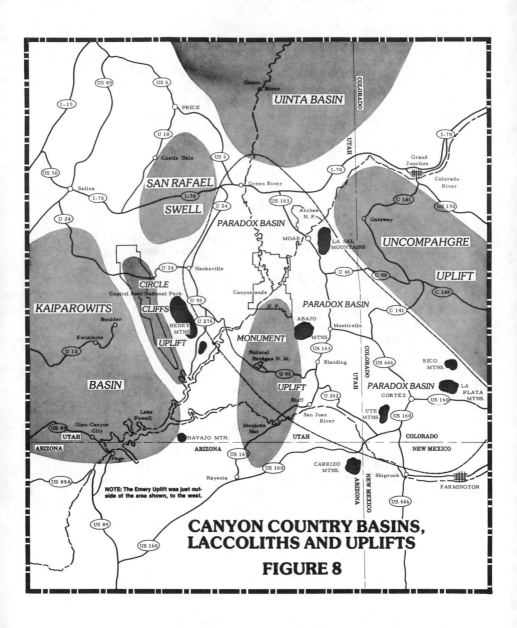

**CANYON COUNTRY BASINS,
LACCOLITHS AND UPLIFTS**

FIGURE 8

MAJOR FEATURES

BASINS. Previous chapters have noted various basins within and immediately adjacent to canyon country, specifically Paradox Basin, Kaiparowits Basin and the Uinta Basin. Of these, only the Uinta Basin to the north of canyon country is now recognizable as a basin. The others have been so distorted by other geologic events since their formation that their previous existence is now obvious only to geologists. The former basins that were within or that affected canyon country geology are shown in Figure 8.

The area that was once the broad lowland of the Kaiparowits Basin is now dominated by the looming Kaiparowits Plateau. The accelerated erosion that began about 10 million years ago reversed the area's topography, leaving rock strata once deep beneath a basin floor now exposed on top of a great plateau. This same reversal has occurred in other former basin areas. The once immense Paradox Basin is now so distorted by mountain forming, uplifts, erosion and the formation of salt valleys that its former outlines have been determined only by the careful study of rock strata. Other ancient basins near canyon country suffered the same fate.

The Uinta Basin still exists, although only as a shadow of its former self. It once extended into the northern reaches of canyon country, where the Book Cliffs now loom high.

In sum, most of the great lowland basins that once dominated canyon country and the Colorado Plateau no longer exist as surface features, and canyon country presently has no recognizable basin-type features.

---CLINES. There are three general types of "---clines." These are "anticline," "syncline," and "monocline." Each of these becomes a specific type of geologic feature, and all are visible many places in canyon country. There are still other types of "---clines," but these are generally of interest only to geologists.

Anticlines. An anticline is an upward arching of rock strata. The resulting bulge may be almost any size, from a few hundred yards to many miles in length, or any shape, from a slender oval to circular. It may even have been buried by later deposits or might even now lie beneath a valley or canyon. The important factor is the upward arching rock strata, not the surface shape, even though the strata may be partly missing.

If an anticline exceeds a certain size or attains a certain shape, it may be called a "swell," or even an "uplift," although not all "uplifts" remain anticlines. There are also several special types of anticlines, but most of these are of interest only to geologists.

There are countless anticlines of various types and sizes within and beneath canyon country. Some are revealed where river gorges cut deeply into rock strata. The strata of one such anticline begin to arch upward just downriver from the end of the Utah 279 pavement, near the potash mill, and continue on downriver for several miles. This anticline is also visible from Cane Creek Road out of Moab, and from the Potash, Chicken Corners and Cane Creek Canyon off-road vehicle trails. The Hurrah Pass trail actually ascends and descends the tilted anticline strata, and the "pass" is the summit of an anticlinal ridge between two parallel canyons cut by the Colorado River and Cane Creek. This conspicuous and readily accessible anticline is also visible from a high mesa-top promontory above Hurrah Pass called Anticline Overlook. This overlook is part of the Canyon Rims Recreation Area. It has a geologic display and is accessible to highway type vehicles.

The San Rafael Swell, in the northwestern corner of canyon country, is a very large anticline that is roughly oval in shape and some 50 miles long on the surface. It is perhaps half again that long beneath the surface. This immense and spectacular anticline or swell had two intervals of upward movement, as noted in earlier chapters, and has had its upper strata severely eroded into a fantasy of colorful shapes. Around its exposed eastern perimeter, harder Navajo Sandstone has been uptilted and eroded into a vast "reef" of jagged spires. The San Rafael Swell is discussed further under "uplifts."

The various salt valleys in the vicinity of the La Sal Mountains are also anticlines, even though valleys on the surface. These are discussed further under "salt valleys."

Synclines. A syncline is a downward arching or sagging of rock strata and is the opposite of an anticline. The resulting lowland may be of any size or shape. Very large synclines may be called "basins." By their nature, synclines are less numerous because anticlines are formed by powerful forces beneath the Earth's crust, while much weaker gravitational forces are primary in the shaping of synclines. When exposed on the surface, anticlines erode from their centers, which were often fractured during their formation, thus exposing their rounded or domed shape. Exposed synclines tend to be obscured from view, because their edges erode down, and their lower centers tend to fill with sediments.

There are many synclines in canyon country, but few exposed and obvious to casual viewing. The best places to watch for synclines are along the major river gorges, where their cross-sections are revealed.

Monoclines. A monocline is where rock strata change elevations abruptly in a stairstep manner, or where they are tilted over a considerable area, with no apparent leveling off. Elevation changes may vary from a few yards to a thousand feet or more, and such stairsteps or areas of tilted strata may be many miles long. The slanted part of the monocline, between its two relatively level "steps" or otherwise, may be almost any angle.

Sometimes the outer edges of very large uplifts or anticlines are called monoclines, especially if the strata on the protruding top of the dome have been removed by erosion, leaving only a "reef" of tilted strata along the anticline's edges.

For example, the eastern edge of the San Rafael Swell could be called a monocline, as could Comb Ridge, along the eastern edge of the Monument Uplift. Waterpocket Fold, along the eastern edge of the Circle Cliffs Uplift, could also be termed a monocline.

There are countless other monoclines within canyon country of various sizes, some revealed along the river gorges, but the most conspicuous are those noted above.

FAULTS AND JOINTS. Both faults and joints are fractures in rock strata, but with a critical difference that is related to how these two types of fractures form.

Faults are fractures in which the rock on one side of the fracture has moved with respect to the other, or is offset to some degree and in some direction along the plane of the fracture. Faults are classified further, depending upon the type of offset, but such classifications are generally of interest only to geologists. Faults form in many ways, but in general result from uneven pressures on local rock strata or the underlying "basement" rock. Thus, faults and fault systems may be short and localized, or may involve the full thickness of the continental or oceanic crust and be hundreds, even thousands, of miles long.

Joints are fractures in which the rock faces on each side of the fracture have not moved or been offset along the plane of the fracture. The faces may have separated a little or a lot, by pulling straight apart, and still be classified as joints. Such separations may or may not have later filled with newer sediments, or the faces become re-cemented together again. Joints are created many ways, but in general occur when rock strata are flexed, with little if any pressure difference in local areas, so that offsetting does not occur. Jointing is common where harder rock layers have been flexed by upward bulging anticlines of all sizes, or by sagging synclines. It also occurs at the bends of step-type monoclines.

Certain types of geologic maps show faults as black lines, sometimes coded to indicate directions of offset. Such maps of canyon country show thousands of fault lines, particularly in the interior areas of massive uplifts, where severe crustal distortions have caused more than simple flexing of the rock strata, and along the perimeters of salt valleys. There are also many places where the offsets caused by major faults are clearly visible on the ground.

Joint, in White Rim Sandstone.

One such fault is in the vicinity of the entrance to Arches National Park, where U.S. 163 travels virtually on top of a major fault line, between the highway turnoff into the park and a point about three miles north. For another eight miles, the fault still remains closely parallel to the highway to the west. This major fault has a vertical offset of about fifteen hundred feet. As a result, the Navajo Sandstone that is at road level to the east of the highway near the entrance to Arches, is out of sight above the Wingate Sandstone cliffs to the west.

This same massive slippage has created an unusual coincidence not far north of the U.S. 163/Utah 313 junction, where painted desert blue-gray Morrison hills west of the highway line up with the much older gray-hued Chinle Formation. To the casual observer, these two strata may seem to be the same one, but the fault offset that separates them is clearly visible in the western bluff, where the cliff face changes abruptly from sheer slickrock to softer sediments.

Joints are extremely plentiful and obvious in canyon country. Almost anywhere that harder sandstone layers have been exposed by erosion, a pattern of fracture lines can be seen. Such lines are usually emphasized by increased erosion along the lines, giving large areas of weathered sandstone the appearance of gigantic biscuits baked close together. Such closed jointing is conspicuous along the White Rim, in Canyonlands National Park, and countless other areas where water deposited sandstone "slickrock" is exposed.

Joints that have separated are also visible in some areas of canyon country, such as the Needles District of Canyonlands National Park. Here, the dominant Cedar Mesa Sandstone has been profusely jointed, with many of the joints separated, and others opened by erosion. Aerial views of this area reveal the startling appearance and extent of an area shaped by jointing, with complications introduced by faulting. The original jointing was caused by the rising Monument Uplift, but relatively recent faulting and still more jointing have added fracture patterns that cross the original uplift patterns, creating the cross-hatched fault/joint basis for the novel beauty of the Needles District.

While aerial views reveal the complexity of this unique fault/joint pattern, a close view is afforded from the many hiking and off-road vehicle trails in the district. Among these, the Elephant Hill off-road vehicle trail and the Joint Trail hiking trail are outstanding.

FOLDS. Fold is a very general term that geologists apply to almost any bending of rock strata, whatever the cause or overall configuration. There are many special types of folds, but these classifications are of interest only to geologists. Sometimes the term "fold" becomes attached to a particular folded structure. Waterpocket Fold, in western canyon country, for example. Here, the eastern edge of the Circle Cliffs Uplift, or anticline, can also be called a monocline, but has been given the formal name of "Waterpocket Fold" to the south, and "Capitol Reef" within that national park. Another ridge of uplift strata farther east is called Caineville Reef. This application of several terms — uplift, anticline, monocline, fold and reef — to the same geologic feature can be confusing to the casual observer.

LACCOLITHS. "Laccolith" is a scientific term synthesized from Greek word components. Translated, it means "stone cistern," which refers to the underground basins or lakes of igneous rock which are typical of laccolithic-type mountains.

Earlier chapters indicated when canyon country laccolithic mountains formed and what the motivating forces probably were. How they form is equally fascinating, although only a general outline of the process can be given here.

Within the Four Corners region of the Colorado Plateau, there were eight major laccolithic intrusions. These are The Henry, La Sal, Abajo and Navajo mountains in Utah; the Rico, Ute and La Plata mountains in Colorado, and the Carrizo mountains in Arizona. The locations of these laccolithic mountains are shown on Figure 8. Of the eight, only the four Utah laccoliths are within the canyon country region covered by this book.

All laccolithic mountains form by the same process, with minor variations introduced by the thickness and type of overlying rock strata and other factors. The process begins as molten, sub-crustal rock forces its way upward along crustal faults. As this magma reaches the softer sedimentary rock strata, it continues upward in a more or less vertical column, or "stock." Where the hot, flowing magma finds weaknesses between the sedimentary strata, it fans out horizontally between strata, sometimes bulging the overlying strata upward into a flattened dome.

Such gigantic, upward-bulging, leaf-shaped, roughly level intrusions of magma are called "laccoliths." A single stock may form many laccoliths of various sizes and shapes, and at various levels, much like a gigantic tree sprouting immense, spatulate, horizontal leaves along its trunk.

If the vertical stock could have broken through on the surface, the result would have been a volcano, perhaps with a few thin, lateral "sills," or horizontal, flat intrusions between rock strata. But the exceptionally thick Colorado Plateau continental crust and overlying strata prevented this in the Four Corners vicinity, so magma stocks there spent their energies creating more sills and fattening them into laccoliths.

As each parent stock and its family of laccoliths grew upward, the overlying sedimentary rock layers were also thrust upward, with the uppermost strata terribly fractured and broken. Even so, after the initial formation of a laccolithic mountain, or a range from several stocks, the upper surface more closely resembled high, rounded hills, than the rugged peaks visible today.

The present laccolithic mountains of canyon country were essentially just such rounded highlands until the general regional uplift 10 million years ago drastically accelerated erosion processes. Water erosion was aided by glacial cutting during the two ice ages that lightly touched canyon country. Together, water and glacial erosion exposed some of the cooled magma, or igneous rock, that formed the laccolithic mountains, and carved their spectacular present shapes from the original rounded hills.

Laccolithic mountains vary in size and the amount of igneous rock presently exposed. Some ranges, such as the La Sals, Abajos and Henrys, have several parent stocks and a complex of laccoliths that have formed the multiple peaks presently seen, with igneous rock from their stocks and laccoliths exposed by erosion on their peaks and lower slopes. Navajo Mountain is believed to have just one stock and branching laccoliths, all of them still deeply buried beneath domed sedimentary strata, with no igneous rock exposed.

To the casual observer, laccolithic mountains appear to be just like others, but closer examination reveals differences. These differences make canyon country mountains unique, for their beauty as well as geology. The "foothills" of all four canyon country laccolithic ranges consist of ancient sedimentary strata, slanting upward toward the mountains and cut deeply with complex canyons. This combination of colorful, tilted foothill strata, eroded into canyons by the flow of mountain water, has played a major part in shaping canyon country as it is today.

Other laccolithic factors also contribute toward the unique beauty of canyon country mountains. Where the higher igneous material is exposed, it fractures into great conical peaks of gray-hued boulders and rocks that form steep slopes. The instability of these slopes, plus their elevation, makes them difficult for vegetation to conquer. Thus, the main peaks of the La Sals are spectacular, barren cones of loose rubble that are usually snow-capped for half of each year.

The Henrys, Abajos and Navajo Mountain also display another type of unique beauty. Monstrous, tilted slabs of red and white sandstone lie on their lower slopes many places, bearing mute testimony to the powerful forces that formed the mountains and deformed ancient canyon country rock strata. In the Abajos, such slabs are readily apparent from the air in several areas. They are spectacular around the base of Navajo Mountain when viewed from the air, or from Lake Powell in the vicinity of Oak Creek Bay. In the Henrys, such tilted slabs are clearly visible around the base of Mount Holmes, when viewed from Utah 276 as it travels through the range on its way to the Bullfrog development on Lake Powell.

It is interesting to note that virtually none of the original sedimentary rock remains in the higher elevations of the La Sal Mountains. It has largely eroded away, exposing the much younger igneous rock. Even more of the laccolith-forming magma, now cooled into igneous rock, is presently exposed in the Henry Mountains. In the other extreme, none is exposed on Navajo Mountain, although scientific tests indicate it is there, still deeply buried beneath distorted

Fisher Mesa.

sedimentary strata.

PLATEAUS. "Plateau" is synonymous with the Spanish word "mesa," a term which is widely used in the southwestern states. Both terms mean a highland that is fairly flat on top, with one or more of its sides being a steep cliff. In canyon country the two words are used interchangeably to describe a common but spectacular land form.

This land form is the result of several factors, including the mechanical strength of various types of rock, their erosional characteristics and thickness, and the alternate hard-soft layering that is common in canyon country sedimentary deposits. Thick, hard layers of sandstone often stand on soft bases and, when undercut by erosion, scale away to form sheer cliffs. The upper surfaces of some moderately durable rock strata are often protected from rain erosion by harder layers, and long, undulating clifflines are produced.

Water courses are another major factor in the formation of plateaus. Their wandering meanders often isolate areas of land, turning them into sheer or steep-walled plateaus.

Canyon country has plateaus or mesas of every size and shape. The immense Colorado Plateau, itself, has distinct, steep-sided edges in places where its perimeter has not been obscured by mountain-building. The Mogollon Rim in Arizona is one such edge. East-central canyon country, between Monticello and the La Sal Mountains, is dominated by a series of gigantic step-like plateaus that descend downward toward the west, to end at the Colorado River. The meandering western rim of the Canyon Rims Recreation Area, where it looms high above the Needles District of Canyonlands National Park, is one such step, or plateau.

Smaller, more distinct plateaus surround the Abajo and La Sal mountains, where river and stream courses flowing from the mountains have isolated higher, stratified land between deep canyons. Fisher Mesa, between Professor Valley and Fisher Valley is an outstanding example of this type of plateau. It is visible from Utah 128, the Castleton-Gateway road and several off-road vehicle trails in the vicinity.

To the west of Moab, the converging Green and Colorado river gorges and their tributary canyons have formed an immense, sheer-walled plateau that narrows as the rivers grow nearer. A final slender peninsula of this high plateau becomes the "Island in the Sky" of Canyonlands National Park, one of the most complex and spectacular plateaus anywhere. Dead Horse Point, in the state park with that name, is another slender finger of the same gigantic, river-carved plateau. Farther south, Junction Butte is a detached tip of the plateau.

Within other areas of canyon country, there are many more examples of various types of plateaus, all born of ancient sedimentary strata and shaped by megayears of relentless erosion.

RIVER GORGES. All the rivers of canyon country lie within deep, winding gorges. As noted earlier, there were no rivers or perennial streams in canyon country earlier than 10 million years ago, when the great regional uplift began. This uplift increased precipitation on the highlands to the north and east of the region, and this water gradually created a network of rivers that were fed by perennial or seasonal streams, or flash-flooding drywashes.

At first, the land was gently sloped, and the water courses meandered. As the land rose and the water flow increased, the meanders were cut deeper. Two ice ages brought still more water and more cutting. The mechanics of river and stream cutting and the formation of "entrenched meanders" is quite complex, but the principal factors in the formation of the magnificent river gorges of canyon country were the kinds of sedimentary rock involved, the fairly rapid rising of the region, and the slow but steady increase of water flow.

The shape of a river gorge at any one place is largely determined by the rock strata that make up its walls. Thick, solid layers of sandstone create sheer walls, such as those along Utah 279, downriver from Moab Valley. Alternating hard-soft layers create terraced set-back walls. There are many places along the San Juan River where the gorge walls are like this. One stretch can be viewed from the Goosenecks of the San Juan Overlook, near Mexican Hat.

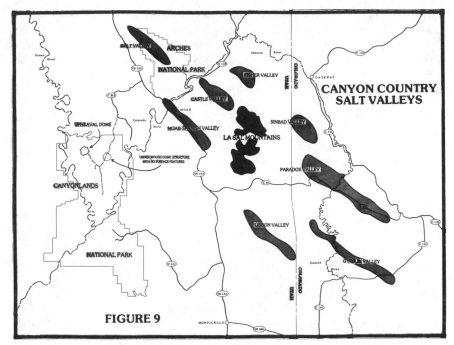

FIGURE 9

The best way to see a wide variety of canyon country river gorge is to take a rubber raft trip down the Colorado, Green or San Juan river, although a good sample can be seen by driving the lengths of the only two highways that travel the Colorado river gorge anywhere, Utah 128 and Utah 279, upriver and downriver from Moab Valley.

SALT VALLEYS. The history of canyon country's eight long salt valleys began early in the Pennsylvanian period, some 320 million years ago, as the region was once again invaded by the western sea. Soon, a large area in what is now eastern Utah and western Colorado began to subside, creating a great, shallow basin with limited inlets. The subsidence caused faulting in the basin bottom, and some of the blocks of rock tilted, creating underground ridges.

Over millions of years, this basin settled slowly and accumulated deposits of various salts, some by evaporation, some by precipitation. This immense "solar evaporation pond" was constantly renewed by more sea water that entered through narrow inlets. Then, the basin was split in two by the initial rising of the Uncompahgre Uplift. This led to a long interval of time during which the western half-basin that was in canyon country, called the Paradox Basin, subsided still more and accumulated more salts, plus sediments from the Uncompahgre, while the Uncompahgre grew spasmodically.

When the sea eventually retreated, sediments from the Uncompahgre completely buried the older salt deposits to great depths. The weight of all this rock overburden started the salt flowing, in a

96

Upheaval Dome, Canyonlands National Park.

"plastic" slow-motion manner, away from the greatest weight near the base of the Uncompahgre and toward the southwest. As it flowed, the salt encountered barriers, the underground ridges of the old fault-blocks of rock that had tilted earlier. These ridges forced the flowing salt upward through the rock strata above the ridges, deforming them into arching anticlines, but not actually reaching the surface. Within these anticlines, Paradox Formation salts are very deep, as noted earlier. It should be remembered that the Paradox Formation salts "flowed" at a geologic rate indiscernible on the human time scale, taking some 150 million years to reach where they are today.

Some deformation of the Paradox Basin salt layers took place at later intervals, such as each time the Uncompahgre rose still higher, or when the La Sal Mountains laccoliths forced their way upward through some of the same fault zones that helped shape the salt anticline. One small igneous intrusion even shouldered up through a small earlier salt dome. This is now exposed in Castle Valley near Moab and is called Round Mountain.

The next step in the formation of the salt valleys began 10 million years ago, with the great regional uplift that triggered the erosional cycle that shaped the present canyon country. As this erosion created the present rivers and streams, entrenched them deeply and removed a mile or so of rock strata from canyon country, it also shaped the present salt valleys, each one directly over one of the long-buried salt anticlines.

As the rivers and streams entrenched and eventually reached the level of the tops of the salts in the long, slender salt anticlines, water slowly dissolved out the more soluble salts, leaving the less soluble gypsum and other minerals and interlayered sediments. The dissolved salts drained away into the rivers, leaving underground gaps which were promptly filled by settling of the rock strata above. This process slowly created the eight present salt valleys that lie to the southwest of the Uncompahgre, parallel to it but punctured in their midst by the La Sal Mountains.

Further surface erosion also worked to deepen the valleys and, in some places, even exposed parts of the upper Paradox Formation. See the earlier chapter on rock strata for some sample exposures. The eight salt valleys that surround the La Sals are Moab, Castle, Fisher, Lisbon and Salt valleys in Utah, and Sinbad, Gypsum and Paradox valleys in Colorado. Actually, Salt, Fisher, and Sinbad valleys are all parts of the same long, slender salt anticline, as are Castle and Paradox valleys. One group of La Sal Mountain laccoliths intruded through the Castle-Paradox anticline, and another group distorted the southeastern end of the Moab Valley anticline.

Today, eight slender, unique valleys lie above the ridges of ancient salt anticlines, where deposits left by a Pennsylvanian period sea were forced upward through and under newer strata, demonstrating how a depressed, valley-like surface feature can lie above an upward arching anticline. See Figure 9. Moab Valley provides ample clues to its underlying anticline, especially along its southwestern cliffline, where the rock strata are visibly tilted upward along the valley's edge. This tilt is readily apparent where the Colorado River leaves Moab Valley, paralleled by Utah 279 on one side and Cane Creek Road on the other. The first mile of the Moab Rim off-road vehicle trail ascends these steeply tilted strata.

It should be noted here that there are still other salt anticlines and domes within the Paradox Basin area of canyon country. Some geologists contend that Upheaval Dome, in Canyonlands National Park, is a salt dome. Others propose different hypotheses to explain this spectacular surface feature. To date, none of the hypotheses is supported by conclusive evidence. Although National Park Service literature favors the salt dome explanation, there is evidence that this is not the full story. Although it does not show on the surface, magnetic surveys reveal that there is a similar mysterious structure below the surface in the Grays Pasture area, about five miles east of Upheaval Dome.

UPLIFTS. As mentioned in earlier chapters, over the ages there have been five major uplifts within canyon country or on its borders. The remnants of four of these are still visible, and contribute to the geologic variety and scenic beauty of the region. The fifth has been lost to ordinary viewing for at least 250 million years although geologists can still find traces of its earlier existence. Following are some details about the five canyon country uplifts. See Figure 8 for their approximate locations.

As noted earlier, the upper, rounded surfaces of anticline-type uplifts are always severely fractured during the uplifting. This promotes an erosion pattern that removes the fractured dome of the uplift, exposes the upthrust older strata beneath that dome, and leaves a monocline of tilted strata around the perimeter of the uplift. All four of the remaining canyon country uplifts followed this pattern, with minor variations.

Emery Uplift (Paiute Platform). This was one of the earliest major uplifts to affect canyon country. It thrust upward in what is now south-central Utah during the early Pennsylvanian period, about the same time as the Ancestral Rockies were forming in central Colorado, about 320 million years ago. For the next 60 million years it contributed sediments to canyon country and acted as a barrier to invading seas. During the early Permian period, the Emery Uplift was finally buried deeply beneath new sediments. Subsequent deposits, uplifts and other distortions of the area, plus millions of years of erosion, have left little obvious trace of the ancient Emery Uplift, even though while it existed it did its share toward shaping the canyon country of today.

Uncompahgre Uplift. This early uplift first started about when the Emery Uplift formed, and was a western sub-range of the Ancestral Rockies, but it continued to grow over the ages. For one long interval, the Uncompahgre highlands were worn down and covered over completely by new sediments, but it later climbed to new heights.

The Uncompahgre had eight intervals of uplift. Their approximate ages were 310, 295, 225, 210, 65, 38, 10 and 1 million years ago. The intervals of uplift growth can be determined from the ages of the sediments that were bent upward each time along its flanks.

Throughout its long life, the mighty Uncompahgre contributed countless cubic miles of sediments toward the formation of canyon country. It also helped shape the ancient Paradox Basin and its many salt valleys, and influenced erosion patterns in many ways. Indeed, it is safe to say that the Uncompahgre Uplift was probably the single most important factor in the creation of canyon country as it is today except, perhaps, for the general regional uplift about 10 million years ago that started the current erosional cycle.

Today, the Uncompahgre still exists as a prominent feature, although only as a shadow of its former self. Its present highlands angle southeastward, just east of the Colorado-Utah boundary, but from high viewpoints its influence on eastern Utah rock strata is apparent. The Dolores and San Miguel rivers mark its steeper southwestern flank, and many backcountry roads and trails penetrate its lovely, wooded highlands. Its ancient Precambrian rock spine is cut by Unaweep

Canyon, along Colorado 141 between Gateway and Grand Junction, Colorado, and by the Colorado River in Westwater Canyon, in eastern Utah.

Ordinary maps now call this immense, long-lived uplift the Uncompahgre Plateau, but geologic maps still call it an uplift, and geologists recognize that its life is far from over yet.

Circle Cliffs Uplift. This uplift formed during the Laramide Orogeny, the same 30 million-year interval of geologic violence that thrust up the present Rocky Mountains and two other canyon country uplifts. For the cause of this violence, see the earlier chapter on Events, Causes and Effects. This uplift rose just west of the present Henry Mountains, but before they formed.

Today, the uplift is exposed as a long monocline of rock strata that is called Waterpocket Fold, because of the thousands of shallow potholes in its high slickrock spine that fill with water when it rains. To the west of this colorful sandstone spine, erosion has shaped the Circle Cliffs. To the north, one extension of the monocline is called Caineville Reef, while another part of the uplift is called Capitol Reef.

Much of the spectacular eastern monocline of the Circle Cliffs Uplift is now included in Capitol Reef National Park. Utah 24 parallels the Fremont River as it cuts through the "reef," and several back-country roads and trails to the north and south offer other access to this scenic geologic highlight. The southern tip of Waterpocket Fold can be viewed from the Bullfrog Basin and Halls Creek Bay areas of Lake Powell.

Monument Uplift. This great oval uplift is quite large and lies in the very heart of canyon country. As exposed by the last 10 million years of erosion, this north-south uplift extends from near the U.S. 160-U.S. 163 highway junction in Arizona, to the vicinity of the Green-Colorado river confluence. The overall outlines of an uplift this size can only be grasped from a high aerial viewpoint, but there are several places where parts of the Monument Uplift are conspicuous highlights of the canyon country scene.

Between Kayenta, Arizona, and Bluff, Utah, U.S. 163 angles across the southern tip of the uplift. There, massive jointing caused by the uplift has helped shape the harder de Chelly Sandstone into a wonderland of soaring spires, buttes and plateaus called Monument Valley. Between Bluff and Lake Powell to the west, the San Juan River cuts deeply across the Monument Uplift, exposing some of the oldest rock in canyon country. San Juan river runners get an excellent cross-sectional view of uplift rock strata.

Just west of Bluff, U.S. 163 cuts through Comb Ridge, the great slickrock monocline that marks the southeastern edge of the Monument Uplift. Utah 95 farther north makes a spectacular passage through Comb Ridge to the west of Blanding, then for the next fifty miles or so crosses the eroded upper surface of the uplift. The wooded

Monument Uplift northern end, Needles District,Canyonlands National Park.

highlands to the west of the Abajo Mountain peaks are a part of the arching uplift. The Abajos intruded into the eastern flank of the uplift after it formed, and somewhat obscured its outlines in their vicinity.

North of the mountains, the uplift slants downward and disappears from view. As noted earlier, this rounded northern tip of the uplift caused the massive jointing that ultimately eroded into the spectacular geologic forms that make the Needles District of Canyonlands National Park such a scenic highlight.

It is interesting to note that over much of the Monument Uplift interior, downward erosion has been considerably slowed when it reached the harder Cedar Mesa Sandstone, which is now exposed over vast areas of the uplift. This sandstone layer, deeply eroded and cut by river gorges, is the red-banded white rock that dominates both the Maze and Needles districts of Canyonlands National Park, as well as Natural Bridges National Monument and vast areas south of the Abajo Mountains. Utah 95 travels across this exposed sandstone layer from just west of U.S. 163 to the Hite area of Lake Powell.

Although the Uncompahgre Uplift is first when it comes to the forming and shaping of canyon country, the Monument Uplift runs a close second. The Uncompahgre contributed both sediments and shaping, while the more short-lived Monument Uplift primarily shaped, but the resulting rock sculpting is outstandingly beautiful.

San Rafael Swell. The San Rafael Swell is a two-stage anticline-type uplift. It first bulged upward during the Laramide Orogeny that worked its violence on the present mountain states between 50 and 80 million years ago. A second San Rafael uplift occurred around 38 million years ago, about when the North American continent first nudged into the Pacific plate. See the earlier chapter on continental drift for more details about this encounter.

The San Rafael Swell is a giant oval that is roughly outlined by U.S. 6 and Utah 24 on the north, east and south, and by Utah 10 on the west. Interstate 70 cuts across its center, and offers excellent views of its eastern monocline edge or reef, and its raised, eroded interior. Toward the west, as I-70 leaves the uplift, the colorful bands and shapes of the rock formations within the uplift are exposed to easy viewing.

Utah 24 travels closely parallel to the eastern monocline of the uplift, known as the San Rafael Reef, providing a continuous view of this spectacular feature for many miles. U.S. 6 offers views of the northern reef, while Utah 10 travels the less conspicuous contours of the western edge of the uplift, where its tilted strata angle gently downward. A number of roads and off-road vehicle trails penetrate the uplift interior, some from I-70 exits, some from the perimeter roads, but only an air tour can reveal the overall magnitude and beauty of this immense canyon country uplift.

It is interesting to note that three of the four canyon country uplifts, the San Rafael, Circle Cliffs and Monument, have steeply tilted monoclines on their eastern flanks and gentle slopes to the west. Geologists are at a loss to explain this regional pattern. There may also be a relationship between the Circle Cliffs and San Rafael uplifts, because their eastern monoclines tend to phase into each other in the Caineville Mesa vicinity, near Utah 24.

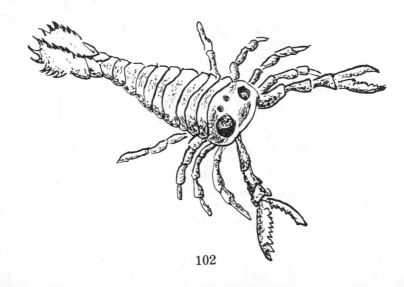

OTHER SURFACE
FEATURES

GENERAL

This chapter describes some of the many other geologic features found
on the surface in canyon country. As with the major features, if there
are only a few of the types of features being described, all of them are
listed, otherwise only representative examples are noted. As an aid to
finding particular feature descriptions, they are listed in alphabetical
order within this chapter, and again in a combination index in the back
of the book.

OTHER FEATURES

Alluvial cuts. Gentle, meandering streams tend to change their courses periodically as they travel the lengths of their valleys or canyons, thus gradually depositing thick layers of sediments. More vigorous streams and sporadic runoff tend to cut into valley and canyon bottoms. One major factor that controls whether a stream builds up sediments, or alluvial deposits, is climate. Another is the amount and type of vegetation on the stream's watershed, or the area it drains. Plant root systems and humus tend to hold water, releasing it slowly in the form of seepage. This creates a gentle, depositing stream. When a watershed lacks vegetation, or loses it somehow, water runs off much more rapidly, creating a violent, cutting stream.

In canyon country, there are countless valleys and canyons where natural conditions have created gentle perennial or seasonal streams, and these in turn have slowly filled rock-walled canyons and valleys with sediments. Vegetation has thrived in these relatively flat canyon floors, with water-loving plants and trees following the slowly shifting routes of the streams. This natural process is an example of sediment building, in the midst of a major geologic interval of erosion.

Human activities, however, are affecting this alluviation process. Heavy grazing, by sheep and cattle, in semi-arid desert areas has severely damaged countless watersheds. The indiscriminate and widespread use of bulldozers on delicate desert terrain, in connection with mineral search activities, has also contributed to watershed destruction. So has the clearing of land for agriculture, and the widespread practice of "chaining" pinyon-juniper forestland by various federal and state land administration agencies. "Chaining" means destroying all the trees and large shrubs in huge areas by dragging heavy anchor chains between giant bulldozers. Hundreds of thousands of acres of wooded canyon country have been chained over the last several decades.

The massive destruction of watersheds by human activities has caused countless canyon country streams to stop depositing sediments and to start cutting deeply into the alluvial deposits they had so patiently built up under natural conditions. Thus, it is fairly common to find valleys and canyons with their alluvial fills cut deeply by their streams and tributary washes, leaving steep banks of soft sediments, and former stream-side trees and shrubs stranded high and dry.

This human-caused erosion can be observed hundreds of places throughout canyon country, and serves to demonstrate how technological man affects geologic processes. Some major examples are Salt Wash, above Wolfe Cabin in Arches National Park; Montezuma Canyon, south of Monticello; Grand Gulch, south of Natural Bridges National Monument; and upper Salt Creek Canyon, in Canyonlands National Park. There are hundreds of other major valleys and washes, and thousands of smaller canyons, in which alluvial cutting can be seen. Although the cutting may be natural in a few of these, most can be traced to human activities, either prehistoric, historic, or current. More details about accelerated alluvial cutting can be found in many scientific papers, but are scarce in popular literature because the subject is politically unpopular.

Ancestral Lake. Several thousand years ago, flash flooding down one tributary of Lake Canyon, in the Halls Crossing vicinity of Lake Powell, slowly dammed another tributary with debris and sediments. Water collected behind this natural dam and plant life added organic material to help seal it from leakage. Eventually, the flooding tributary overflowed its own canyon and its waters cut a new channel into the reservoir it had created. This bigger supply of water filled the reservoir, creating Lake Pagahrit, the name given this tiny desert lake by the Paiute Indians. Later, Mormon settlers named it Hermit Lake, and the present Lake Canyon, a tributary of Lake Powell, was named after the freakish desert lake it contained.

Prehistoric Indians also used the lake. The Anasazis lived beside it from about 1050 to 1275 AD, leaving rock structures as evidence of their occupation. At its maximum, Lake Pagahrit was about one-half mile long, one-quarter mile wide and fifty feet deep.

On November 1, 1915, Lake Pagahrit was destroyed by the same natural mechanism that created it, flash flooding. A monstrous flash flood overtopped the natural dam, cut into it and drained the lake. The site of this "ancestral lake" can be reached by hiking up Lake Canyon beyond where Lake Powell has penetrated. More details about this strange lake can be found in one of the books listed under Further Reading, "Canyonlands Country," a collection of scientific reports by members of the Four Corners Geological Society, and in several older reports by historian Albert R. Lyman.

Ancestral riverbed. A deep canyon cuts through the Precambrian granite spine of the ancient Uncompahgre Uplift to the northeast of Gateway, Colorado. Colorado 141 travels this canyon for many miles. Until the late Pliocene epoch, 3 or 4 million years ago, the combined waters of the Colorado and Gunnison rivers flowed through Unaweep Canyon, slowly cutting into the bedrock of the ancient uplift.

Then, by an erosional process called "stream piracy," the Colorado river was diverted into its present course around the northern end of the Uncompahgre, leaving the Gunnison still flowing through Unaweep Canyon. A second act of stream piracy next captured the Gunnison too, leaving Unaweep with only a trickle of water.

Unaweep Canyon, ancestral riverbed of the Colorado River.

About 1 million years ago, the Uncompahgre rose another 1500 feet. This finished shaping Unaweep Canyon as it is today, with West Creek flowing west from the newly elevated center of the long canyon, and East Creek flowing east from the center and into the Gunnison, but far upriver of its present confluence. A final act of stream piracy during the last million years diverted East Creek to its present confluence near the Colorado town of Whitewater.

More details about how Unaweep Canyon was changed from Colorado River gorge to ancestral riverbed can be found in   "The Geologic Story of Colorado National Monument," by S.W. Lohman of the U.S. Geological Survey, listed under Further Reading.

Arches. There are innumerable natural holes in the rocks of canyon country. Many are small and insignificant. Some are larger and more

Corona Arch.

aesthetically pleasing. A few are spectacular in size, shape and beauty.
There is a human tendency to want to classify, label and name the
larger or more interestingly shaped holes.

Unfortunately, no two classifiers of natural rock openings agree
upon the same set of definitions, and the subject is not geologically
important enough to warrant a standardization effort by national or
international organizations of geologists.

For purposes of this book, natural rock openings have been
grouped into four categories: arches, bridges, tunnels and windows.
Defining each of these categories can be very complex. These complexities will be pursued in another canyon country guidebook.

In general, an arch is defined partly on the basis of what it is not.
A natural bridge spans an existing or former watercourse. An arch
does not. A tunnel is an elongated hole through a rock mass. An arch
is not. A window is a hole through a thin wall of rock, so far from any

wall edge that the hole does not give obvious shape to that edge. An arch opening does give shape to that edge.

Arches are created by wind and water erosion, but not flowing-water erosion. Arch openings start in thin walls of rock at points where the chemical composition of the rock allows faster erosion by rain and seeping moisture. In a very few cases, the opening may be started by a rock collapse based upon the release of internal rock strain. Once the opening starts, it is slowly enlarged and rounded by a combination of erosional and internal stress forces.

Water is the main erosional force. Wind is minor, because most arches are too high above the ground for abrasion by wind-borne sand. Moisture uses both chemical and mechanical methods to enlarge arch openings. It dissolves the chemical cements that bind the rock particles, allowing them to slough away. It freezes and expands, fracturing the rock. In sum, when it comes to shaping arches, water does the work, and wind cleans up the mess, by blowing away the loosened rock particles.

Scientists believe most present arches have been created within the last million years, with high-moisture glacial intervals accelerating arch formation and shaping. Even today, however, the process continues. New arches are started, young arches are shaped and aged by erosion, and old arches collapse.

Unfortunately, human activities threaten the premature collapse of a few of the older spans. Canyon country residents at one time shuddered with each sonic boom produced by overflying supersonic military aircraft, fearful that the blast of sound might destroy ancient and delicate Landscape Arch in Arches National Park. Such overflights have now been curtailed, but the explosive blasts of mineral-search activities in the vicinity of the park still present a hazard.

Such awesome spans as Landscape Arch have only survived as long as they have because the Colorado Plateau, and canyon country, are relatively free of earthquakes. The thicker Colorado Plateau crust protects canyon country spans, and balanced rocks, too. Very minor tremors occur in the region with fair regularity but noticeable earthquakes have not disturbed the region for a long time, and are not likely to in the future.

Arches tend to occur more often in certain types of rock, although they can occur in a wide variety of rock. In canyon country, arches generally form in the thicker, more uniform sandstone strata, such as wind-deposited Wingate, Navajo and Entrada, or such shoreline deposits as Cedar Mesa. Most of the arches in Arches National Park occur in Entrada sandstone, while most of the arches in Canyonlands National Park are in Cedar Mesa sandstone. Arch-type openings also occur in many other geologic rock strata, both within and outside of the region's national park areas.

Balanced rocks. Balanced rocks are nature's successful gambles, its long-shots that paid off. As erosion wears away at exposed rock strata,

Happy Turk balanced rock, Cane Creek Canyon.

there are countless chances for the erosion to be so uniform around a particular chunk of rock that it doesn't tip over but remains balanced on a smaller base. The odds against this delicate balancing act are enormous, astronomical. Even so, out of the multi-billions of big rocks and boulders that might have become balanced rocks, by sheer chance a few succeed. But only for a bare instant in time, because erosion continues, inexorably, and sooner or later every balanced rock topples.

In canyon country, conditions are ideal for the creation of balanced rocks, and for their long lives, with humans being a greater hazard to their existence than erosion. The conditions that lead to the creation of balanced rocks in canyon country are: vast areas of exposed rock that has alternating harder and softer layers; the arid climate; the lack of significant earth tremors; and the widespread fracturing of the harder rock layers by faulting and jointing.

The fractured harder layers contribute an endless supply of potential balanced rocks, which in turn protect their pedestals of softer sediments from excessive rain erosion. The arid climate also prevents excessive rain erosion and reduces the growth of vegetation, which can topple a balanced rock with its roots. The absence of tremors permits the creation and relatively long life of balanced rocks. Human activities, such as vandalism, blasting and sonic booms, shorten the all too brief intervals during which balanced rocks defy the laws of gravity and chance.

There are many balanced rocks in canyon country. A few are well known, as in Arches National Park. Others are in the backcountry and less well known. There are many massive, delicately balanced rocks along the White Rim in Canyonlands National Park. Visitors who compare photographs of the Balanced Rock in Arches with the rock itself, will notice that a smaller companion of the main balanced rock is missing. It fell in 1975.

Another picturesque balanced rock in Cane Creek Canyon, near Moab, is called The Happy Turk because, from one angle, it resembles an enormous, turbaned head with a smiling face. This happy monolith is about to meet an unhappy end because its base is perilously eroded, and it is located near an active uranium mine where periodic blasting takes place.

Bridges. Natural bridges are another of the four general types of rock openings. See "Arches." Bridges are cut from either vertical or horizontal sections of relatively thin rock by the action of flowing water, either steady or intermittent. A stream may later change its course and abandon its bridge, but the term still applies.

After running water has started construction of a natural bridge, it may further widen the opening, but other sculpting is done by the same erosional forces that shape arches. The most classically shaped natural bridges have outlines reminiscent of certain types of manmade bridges, flat on top with arching openings. The three spans in Natural Bridges National Monument are this type. Other natural bridges are

Rainbow Bridge, Rainbow Bridge National Monument.

also arched on top. Rainbow Bridge is the ultimate of this type.

Natural bridges tend to form in harder rock strata and, as with arches, can be classified into several types. As with balanced rocks, fractures help in the formation of certain types of natural bridges, such as where water flows over a rim of solid rock, then creates a bridge by diverting through a jointline back from the rim. Mussleman "Arch" in Canyonlands National Park, and the twin Gemini Bridges to the west of Moab, are examples of this type of bridge.

Natural bridges also "age," then collapse. The three main spans in Natural Bridges National Monument are typical of this. There, Kachina is the youngest, with its abutments still being attacked by flowing water erosion. Sipapu is somewhat older. Its abutments are safely above running water erosion, and future shaping will be limited to the slower erosion processes. Owachomo is an old bridge. It has even been abandoned by the stream that originally formed it, and the slower erosional forces have thinned its span. Upcanyon from Sipapu there are remnants of another large bridge that fell long ago.

Canyon streams. There are many perennial, seasonal and intermittent streams in canyon country. The perennial streams usually originate in higher country, then flow downward through lower canyons to the nearest river gorge. Seasonal streams may or may not have higher sources. There are two types of intermittent streams. One type may flow for a few days or weeks following rain or snow-melt, then be dry. The other type may flow steadily or seasonally, but be intermittent on the surface because it flows beneath the sand and gravel of its bed, surfacing only when forced to by solid rock.

All of these types of streams may receive parts of their flows from springs along the canyons. Some may receive all their water that way, and none from a higher source. These springs, in bone-dry desert canyons, are a curious geologic phenomenon.

The sandstone that dominates canyon country is generally fairly porous. It soaks up water from rain and snow, then lets it seep slowly downward, until it encounters an impervious layer. Then the water seeps gradually along the surface of that layer, following the path of least resistence, until it reaches a fracture, a fault, a more porous layer or anything that will let it continue downward.

A lot of this seeping water ultimately joins vast underground water tables or reservoirs, but in canyon country, where the rock strata are so often cut deeply by canyons, water seeping laterally along an impervious layer often reaches a canyon wall. There, the water drips or trickles out into the canyon as a spring. Such springs can appear at any level of a canyon wall, or even along an open cliff line or escarpment. It is quite common in canyon country to see the verdant patch of vegetation that marks a spring-seep high on a sloping bluff, or even in the middle of a massive slickrock wall.

When enough such springs intersect a canyon along its length, a canyon stream is created, making a lush, green linear oasis in the

Negro Bill Canyon.

midst of arid desertland. The water in such streams comes from large areas of the arid highlands above the canyons, and may take years to seep from where it first fell, through miles of slightly porous rock, along a plane of impervious rock to the spring that finally lets it surface and flow, in an isolated desert-canyon stream, far from any obvious source. Such springs and streams are vital to almost all desert wildlife, and enhance the rugged beauty of canyon country.

Areas where canyon streams flow over stretches of gently sloped solid sandstone are sometimes called "slides." Such slides may be very slippery from coatings of water plants, and may be eroded into strange and lovely shapes.

There are many canyon streams in the region. Most must be hiked, because they are too narrow or rugged for vehicles. Some examples are listed in other guide books in this series.

Chicken Corners caves.

Caves. There are two general types of natural caves in canyon country. One type slowly forms when a spring-seep occurs within, or at the base of, a thick, solid wall of sandstone. There, the seeping water weakens the rock, and it slowly scales away, creating a shallow spring-seep cave. Some such caves contain lush vegetation, such as mosses, ferns and other moisture-loving plants. A few become quite deep, but most remain shallow in proportion to their lateral dimensions. Spring-seep caves are common in canyon country wherever there are thick layers of solid sandstone.

The other type of cave is less common. Such caves are created when rainwater penetrates fractures in exposed layers of sandstone, then trickles down and erodes caves below the fractures, especially in the softer sediments below the harder sandstone layer. The formation of such caves requires a combination of circumstances that is not common, so there are relatively few such caves. One group of them can be seen from one spur of the Chicken Corners off-road vehicle trail southwest of Moab. There are a few caves of this type in the Needles District of Canyonlands National Park.

There are also many manmade caves in canyon country, left behind by mineral search activities. These are very hazardous and should not be entered.

Colors of rock. In general, rocks acquire their colors from the minerals they contain. A complete study of rock colors could fill a book but it is appropriate here to provide a few notes about the rock colors that dominate canyon country.

Red is common in much of the region, a color provided by iron oxide, or rust, in the sandstone. The presence of iron in this form indicates the climate was hot and dry when the rock was originally deposited, whether by wind or water.

White or whitish is the color of relatively pure sand. Deposits left by beach sand, beach dunes or off-shore sand bars are generally white or nearly so. Red-hued layers within such sandstone represent land sediments rich in iron oxide washed over the white sand by fresh water streams. Gypsum and limestone layers are sometimes white, but may also be other colors.

Gray or black deposits generally either contain organic matter, or were laid down in stagnant water with little oxygen, or both. Mancos Shale is typical of this type. Grayish layers mixed in with more colorful sandstone layers usually came from mineral-rich volcanic ash that was deposited with other sediments. Painted desert colors, banded grays, reds, blues, greens and purples, come partly from volcanic ash, partly from other minerals.

White or pale green spots or streaks within red-hued sandstone are common within certain water deposited sediments. These are created when the iron oxide in the rock is chemically changed to a form that is not red, but pale green. This form is water soluble and is slowly leached out, leaving white sand showing. The chemical action is usually started by organic matter deposited with the original sediments. Decay products of that matter, perhaps only a tiny seed, cause the chemical reaction.

For the colors of specific canyon country rock layers, refer to the earlier chapter on rock strata. For details about other rock and mineral colors, refer to any of the many good books on mineralogy.

Confluences. In canyon country, the term confluence is commonly applied to the point where two rivers meet, but could also mean where two canyons join, whether the canyons contain rivers or streams, or are dry watercourses. In the more general meaning, there are thousands of confluences within canyon country, many of them quite spectacular, but the most interesting confluences are where the gorges of two rivers meet.

The best known of these is in the heart of Canyonlands National Park, where the Green and Colorado rivers join their sediment-laden waters. This confluence can be seen by air and by boat, but is also accessible by land via a spur of the Elephant Hill off-road vehicle trail in the Needles District of the park.

Of the other major river confluences, the San Juan-Colorado confluence is now a part of Lake Powell, as are the Escalante-Colorado and Dirty Devil-Colorado confluences. The Dolores-Colorado confluence is about one-mile east of the Dewey Bridge, across the Colorado from

Colorado-Green Confluence.

Utah 128, but is rather inconspicuous. The San Rafael-Green confluence is across the river from the ranch at the end of the Floy Wash road, south of the town of Green River, but it is also inconspicuous. The Fremont and Muddy rivers join in open desert terrain, just upstream of the Utah 24 bridge, to become the Dirty Devil River from there on to Lake Powell.

Cross-bedding. In many of the layers of sandstone in canyon country, it is possible to see that the rock is composed of many thin, sloping layers, some quite intricately interfingered or overlapped. In general, such layers were formed when the sand was first deposited, either by water in a stream, or by wind on dunes. As the water flow or wind direction changed, the angles of deposit changed, leaving the cross-bedding that later became a part of the rock.

The various mechanisms of cross-bedding are quite complex, and are explained in detail in most good geology textbooks.

Cryptogamic soil. This term is applied to loose, sandy soil that is held in place by a dark, irregular crust. This crust consists of a complex of microscopic plantlife that binds the soil with its web of rootlets, making the soil highly resistent to wind and water erosion. Without cryptogamic soil, few other forms of life would be able to thrive in the more arid, sandy areas of canyon country.

Cryptogamic soil is common in canyon country and is highly effective as an erosion deterrent. This makes an inconspicuous plant community a vital geologic force. Unfortunately, the crust is very fragile, and grows very slowly. When the crust is broken, this permits wind and water to erode the sandy soil at its normal, rapid rate.

Under natural conditions, few things damage the cryptogamic soil crust except for the small hooves of deer. Human activities, however, are highly destructive. The feet of humans and their domestic animals crush the crust, and human vehicles ruin the fragile layer wholesale, leaving the exposed surface vulnerable to rapid erosion. This is one way in which man acts as a geologic process in canyon country.

Desert varnish. Desert varnish is the dark, lustrous coating often seen on rock surfaces in desert areas. It is composed of either manganese or iron oxides that have been slowly leached from the rock and deposited on its surface.

In canyon country, dark blue-black or brown desert varnish commonly occurs on the surfaces of harder types of sandstone that have been exposed to the air for a long time, and on harder rocks and pebbles in some places. In general, the darker the color, the older the exposure. Prehistoric Indians scratched or chipped through desert varnish layers in making petroglyphs.

There are innumerable exposures of desert varnish throughout canyon country on cliffs and larger boulders, and quite a few places

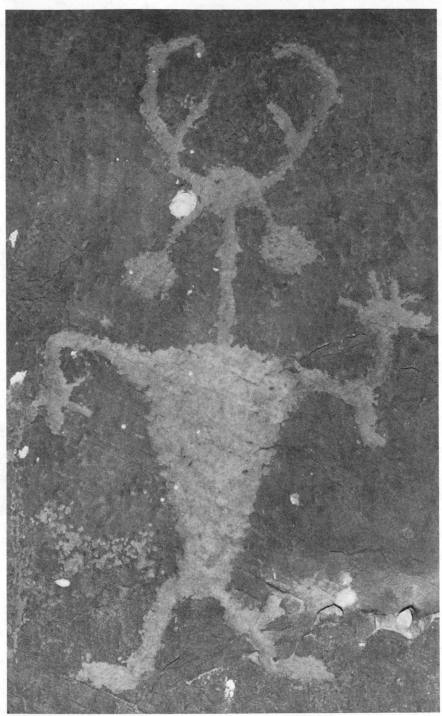
Petroglyph in desert varnish, Colorado River gorge.

where varnish-darkened pebbles can be seen, such as along the Poison Spider Mesa and Chicken Corners off-road vehicle trails near Moab.

**Dikes.** Dikes are thin, more or less vertical walls of igneous rock. They are created by the flow of hot magma into faults and joints. Later erosion may expose the dikes, wearing away the softer, fractured sedimentary rock and leaving thin walls of dark igneous rock standing free. Dikes form in volcanoes, but a few also formed when canyon country's laccolithic mountains were created by intrusive magma, or molten rock.

From Utah 276, several large dikes are visible as huge, jagged fins of black rock near the summit of Mt. Holmes, in the Henry Mountains.

**Dinosaur tracks.** The foot tracks of prehistoric animals occur many places in canyon country. Some were made by the larger reptiles commonly called "dinosaurs." Others were made by smaller reptilian or amphibian species. A few were made by pterosaurs, or flying reptiles.

All of the tracks now found were originally made in mud, then covered by other sediments. Eventually, the mud turned to stone. Many millions of years later, the tracks were uncovered by erosion. The pterosaur tracks found in two locations near Moab were originally made in the mud of a desert playa, or shallow, ephemeral lake. At that time, the region was a vast Sahara-like desert. There are tracks of

Dinosaur track, foot outlined.

several other small animals with the pterosaur tracks.

There is a group of curious dinosaur tracks within 1¼ miles of the center of Moab, but access is difficult and the site is on private land.

At another site within six miles of Moab, the author discovered and reported a number of tracks made by large, three-toed carnivorous dinosaurs. To date, the appropriate land administration agency has not protected the site, and some of the tracks have since been damaged by heavy machinery.

Entrenched meanders. Rivers and streams that cross relatively flat, open country that is gently sloped follow a curving, looping course. They meander. Such rivers and streams quite naturally change courses at intervals, forced out of their beds by accumulated sediments, or by flooding that cuts new channels. As long as it is not interrupted, this process slowly builds up alluvial sediments in the areas where the rivers and streams meander.

When canyon country rivers and streams were first born some 10 million years ago, this region was also relatively flat, open and gently sloped. Water crossing it flowed in curving, looping meanders. But the region changed slowly. It raised in elevation and the water flow increased, slowly but steadily. These changes caused the rivers and streams to cut their meandering courses more deeply, rather than change their courses as normal, to the point where they were trapped

Green River meander.

in their beds. Two glacial ages added extra water flow, causing still more cutting.

As a result of these three factors—regional rise in elevation, gradual increase in water flow, plus extra glacial water—the meandering canyon country rivers and streams entrenched, or deepened, ultimately to create the present curving, meandering canyon complex.

The term "gooseneck" is often applied to a stretch of entrenched river meander so extreme that successive loops almost meet. There are many such gooseneck stretches along the major canyon country rivers. Outstanding stretches can be viewed from Dead Horse Point, above the Colorado River, from Goosenecks of the San Juan Overlook, northwest of Mexican Hat, and from Spring Canyon Point, above the Green River.

Fins. This term is commonly applied to relatively thin slabs of rock that stand upright and have weather-rounded outer contours. Only the thicker, harder sandstone strata can form fins, and then only when they have been subjected to parallel jointing by crustal deformation.

In canyon country, most major groups of sandstone fins have formed where thick, hard strata have been fractured by the formation of anticlines. Later, when the strata were exposed on the surface, erosion slowly widened the joints and rounded the vertical slabs of rock between them, leaving the startling masses of parallel fins visible today.

Fiery Furnace, Arches National Park.

Arches National Park has two such areas of fins, one on each side of the Salt Valley anticline. The mass on the west side is called Klondike Bluffs. On the east side, the Fiery Furnace, Devils Garden and Eagle Park on to the north are in the great row of fins there.

There are also areas of fins in the Needles and Maze districts of Canyonlands National Park, in Capitol Reef National Park, in Glen Canyon National Recreation Area, in Behind-the-Rocks to the west of Moab, and many other locations in canyon country.

Fossils. Fossils are the petrified remains or imprints of prehistoric life forms, either plant or animal. They are fairly common in canyon country. Scientists use certain fossils to help date various geologic strata. This process is described in several of the books listed under Further Reading.

More details about fossil locations appear in the following chapter on rockhounding, but in general the kinds of fossils found in various rock strata depend upon how they were deposited. The fossils in each layer represent the life forms that lived in that vicinity when the sediments accumulated. Marine deposits contain marine life forms. Land deposits contain land life forms. Amphibious life forms may be found in either marine or land deposits. Lakes have their own special marine species. Many marine and lake life forms and fossils are very small, even microscopic in size.

In canyon country, the more obvious and common fossils are sea shells, petrified wood and petrified dinosaur bone, with the rocks in some areas containing the fossilized remains of countless smaller sea creatures.

Gargoyle gardens. Certain canyon country rock strata erode into strange shapes reminiscent of legendary or mythical people or creatures, or beasts out of nightmares. In the western part of the region, erosion has created Goblin Valley out of Entrada Sandstone. In the eastern part, the Cutler Formation erodes into fantasy shapes wherever it is exposed. The Onion Creek off-road vehicle trail goes through one outstanding exposure. There are other areas throughout canyon country where certain rock characteristics and erosional forces have combined to produce "gargoyle gardens."

Grabens. In general, this geologic term is applied to long, narrow sections of surface rock that have sunk between two roughly parallel fault lines, leaving a steep-walled depression that somewhat resembles an old sunken grave, or "graben."

The most spectacular grabens of canyon country are found in the Needles District of Canyonlands National Park. Some have been given such names as Red Lake Canyon, Cyclone Canyon and Devils Lane.

The Needles grabens were produced by a rather complicated process. As the Colorado River, to the northwest of the present grabens, cut downward, the reduced overburden of rock allowed an

Goblin Valley, Goblin Valley State Park.

underlying salt layer, the Paradox Formation, to bulge upward, forming a narrow anticline below the river. When the river ultimately cut down into the top of this salt ridge beneath its bed, this allowed the thick rock strata to the southeast of the river to "stretch" very slightly by "sliding" on the relatively soft layer of salt, in the direction that all the strata tilted, that is, to the northwest, toward the river.

This slight stretch caused parallel faulting, and settling between the faults, thus forming the present grabens. These strange geologic features can be viewed from various spurs of the Elephant Hill off-road vehicle trail in the Needles District, but only an air tour can provide an overall view of this unique area.

Gypsqueeze. Earlier descriptions of anticlines and salt valleys tell how Paradox Formation salts have been forced upward to form various canyon country features. In some places, these salts have been exposed on the surface by erosion. Such exposures usually consist primarily of gypsum, with the more soluble salts long since dissolved and carried away by water. "Gypsqueeze" is a colloquial term sometimes applied to these "squeezed up" masses of whitish mineral.

The Onion Creek off-road vehicle trail east of Moab goes through one such mass in lower Fisher Valley. This particular salt mass has extruded slightly upward in relatively recent times, deforming Pleistocene sedimentary deposits. This flowing was caused by the same removal of overburden that led to the formation of the grabens in Canyonlands National Park. For other "gypsqueeze" exposures, refer to the earlier description of the Paradox Formation.

Gypsum vein, Dolores River gorge.

Gypsum. As noted under "gypsqueeze," masses of Paradox Formation salts exposed on the surface are often largely gypsum, or a hydrated, crystalline form of calcium sulfate. This can also be found in thin layers within other rock strata, such as the Summerville Formation, that contain coastal lagoon deposits. One exposure of such a layer is visible from the Dolores River off-road vehicle trail, downriver from Gateway, Colorado. There are many others throughout canyon country.

Hanging valleys. A hanging valley is one whose floor is appreciably higher than the floor of the valley into which it drains. Most such valleys occur in mountains, where glaciers have deepened main valleys, leaving smaller tributary valleys "hanging," often with their streams making high waterfalls as they plunge into the main valley.

Canyon country also has a few large hanging valleys of a more unusual type. These were formed by such geologic events as faulting and upthrusting anticlines, then were further shaped by erosion. Most of the smaller hanging valleys or canyons in the region were created solely by slower erosion in the tributary canyons. There are thousands of these, and when it rains, they produce spectacular waterfalls in the canyons into which they drain.

Some of the larger hanging valleys, or canyons, that are produced

124

by faulting or anticlines have the peculiar characteristic of draining away from the larger, deeper canyons that they join. This is caused by the rock strata being tilted away from the fault or anticline, with that feature later becoming a lower valley.

Little Canyon, north of Moab, is an excellent example of such a novel hanging valley. One upper end of this long canyon system can be seen from U.S. 163 as a notch or gap in the western cliffline, about 5 miles north of the Colorado River bridge. Little Canyon drains away from this hanging valley junction, eventually to join the Colorado River about 8½ miles downriver from Moab Valley. Several off-road vehicle trails provide access to this unusual canyon at various points.

Jumps. This term is commonly applied to a steep or sheer step-down in a canyon floor. If there is water flowing in the canyon, there is a waterfall at the jump. Jumps often create difficult or impassable barriers to hikers going up or down a canyon. Jumps are fairly common in canyon country canyons, and usually form where a harder rock layer resists erosion in the canyon bottom. Jumps that have water flow only during flash-flooding may also be called "dry waterfalls" the rest of the time.

Two well known jumps are the Upper Jump and Lower Jump of Salt Creek, in the Needles District of Canyonlands National Park. When this seasonal, intermittent creek is flooding from recent rain, the Lower Jump becomes a spectacular waterfall. It can be reached by the Colorado River Overlook off-road vehicle trail in the Needles District.

Lower Jump, Salt Creek, Canyonlands National Park.

Mud crack casts. These odd "fossilized mud cracks" are fairly common in the types of strata that were once mud or silt that had a chance to dry and crack in the familiar patterns. Such mud cracks were later filled by newer sediments, often of a different type and texture. This difference sometimes permitted the rock that the mud became to separate or delaminate at that juncture, or erode selectively, exposing a casting of the original mud cracks, with raised ridges where the cracks were.

Such mud casts are often aesthetically pleasing in pattern, just as the patterns of cracks in drying mud can be, especially after subsequent rain erosion has added texture to the pattern. In canyon country, mud crack casts can be found in several rock strata, but the Moenkopi Formation is a favorite among collectors.

Oil shale. Oil shale is a mixture of organic and mineral sediments that were deposited in a large lake. While there is little oil shale within canyon country, there are large amounts of it immediately to the north in Green River Shale deposits.

Paradoxical valleys. With rare exception, valleys and canyons are higher at one end than the other, and water that flows into a valley, or accumulates there, drains out at the valley's lower end.

Two canyon country valleys are exception to this rule. These great rock-walled valleys are high at both ends, and rivers enter the

Moab Valley river portal.

126

valleys through gaps or "portals" in their rock walls, then leave through portals in their opposite walls. These "paradoxical valleys" are Moab Valley in Utah, crossed by the Colorado River, and Paradox Valley in Colorado, crossed by the Dolores River. Both are salt valleys, as described earlier, and that is the key to their unique nature.

The full story of how these two rivers came to slice directly across these two deep, rock-walled valleys is complex, but can be summarized here.

As described earlier, these two salt valleys did not exist as valleys at the time the rivers were born. The region was relatively flat and open, and the new rivers meandered across plains that revealed no hint of the buried salt anticlines. As the land uplifted, the rivers flowed more vigorously and their meanders were entrenched.

As this cutting and canyon-forming continued over millions of years, thousands of feet of rock were carried away, eventually revealing the uppermost strata that the salt anticlines had distorted. Slowly, the rivers cut down through these strata too, but until the rivers reached the levels of the uppermost salt ridges, the areas that are now valleys were roughly the same elevation as the surrounding land.

Then, as the rivers cut still deeper, the soluble salts were leached out and carried away in the rivers, and subsequent settling formed the valleys. Thus, the salt valleys and river gorges attained their present depths concurrently as two aspects of the same force, river entrenchment, producing by odd coincidence the two paradoxical valleys in eastern canyon country.

Petrified dunes. There are numerous places in canyon country where the rounded, rolling contours of exposed sandstone resemble sand dunes turned magically into rock. Such "petrified dunes" are, indeed, ancient desert sand dunes turned to rock, not by magic but by slow, natural rock-making processes.

Although the petrified dunes effect can occur on the upper surfaces of other desert-dune sandstones, such as Wingate and Entrada, it is common on Navajo Sandstone. Vast expanses of this relatively hard rock layer are exposed throughout canyon country, lending the appearance of dune-scapes immune to the wind.

Navajo Sandstone and other such desert-sand deposits were originally shifting sand dunes with conventional contours, and with the usual laminations of sand caused by changes in wind direction and velocity. When these dunes ultimately were buried by later sediments, the sand grains were slowly cemented together by chemicals brought by seeping water and compacted by the weight of overlying sediments. Thus, sand gradually turned to sandstone.

Millions of years later, as these petrified sand dunes were exposed by erosion, they tended to erode along the rounded planes of the original wind-layered sand laminations, thus resuming some semblance of their original sand-dune shape.

Petrified sand dunes, Sand Flats.

Petrified wood. Petrified wood is fairly common in canyon country, indicating that for various periods of time the region had a climate that encouraged the growth of large trees. Some canyon country petrified wood exposures are in strata so old that the trees were among the earliest to develop.

Contrary to popular belief, petrified wood is not wood that has been entirely replaced by minerals. The softer tissues within the wood are, indeed, replaced by slowly growing, crystalline minerals, but the original tough woody fibers are impregnated by the invading minerals and well-preserved by the surrounding crystalline structure. These fibers can be seen in petrified wood with a strong magnifying glass or microscope.

Petrified Forest National Park in Arizona has great quantities of petrified logs, exposed by erosion of the Chinle Formation. This formation is also present in canyon country. Within the region, petrified wood also occurs in several other rock strata, including the Morrison Formation and the Honaker Trail Formation of the Hermosa Group.

Portals. In canyon country, this term is commonly applied to the openings in high cliff lines through which rivers enter or leave paradoxical valleys, in particular the two river portals in Moab Valley. See the earlier description of "paradoxical valleys" for details on how these strange portals formed.

Petrified log, Little Valley.

Potash. Chemically, potash is potassium carbonate. It has many industrial uses and is thus a valuable mineral. Potash occurs in the Paradox Formation and is produced in canyon country by an unusual process called "solution mining." Water is pumped into an underground potash layer, where it dissolves the soluble mineral. It is then pumped up for concentration by evaporation in ponds on the surface. After concentration, the potash is refined further in a mill. Utah 279, the paved highway that goes downriver from Moab Valley, ends at a potash mill. The Potash off-road vehicle trail that continues beyond the end of Utah 279 passes a wide expanse of potash evaporation ponds. These ponds are also visible from Dead Horse Point. Another mill and other ponds are being planned for an area not far north of Canyonlands National Park and just west of Arches National Park.

Potholes. In canyon country, this term is applied to holes or depressions of varying depths in the upper surfaces of sandstone strata. Such potholes can vary in size from a shallow depression an inch or so deep and a foot or two in diameter, to sheer-walled holes twenty or thirty feet deep and fifty feet or more in diameter.

Potholes of all sizes tend to collect and hold water. In arid regions such as canyon country, this water provides moisture for a variety of life forms, some of them unique to desert regions. Even the shallower potholes, with their rare and brief collections of water, play a vital part in the life cycles of such tiny creatures as fairy shrimp and desert shrimp. Larger potholes may collect sediments, then support their own tiny communities of plants and animals. Some of the larger potholes even develop soils that permit the growth of larger shrubs and even trees. A few that are on drainage lines contain permanent ponds of water, and may develop small colonies of water-loving plants and animals.

Sandstone potholes form by a combination of water and wind erosion. As water from rain or melting snow stands in even the slightest depression on a sandstone surface, the water weakens and dissolves the chemical cements that bind the sand particles. This loosened sand is later washed or blown away, making the depression slightly deeper. The cycle is repeated, and the pothole deepened, until it is interrupted.

This can be done by water erosion cutting the edge of the depression so deeply that it can no longer hold water. Because this type of lateral cutting destroys or limits many potholes that start on gentle slopes, it is not uncommon to find the deepest, surviving potholes in the very tops of sandstone domes, or petrified dunes. The pothole-forming process can also be slowed considerably by the pothole becoming deep enough to hold sediments and vegetation.

Most deeper potholes contain loose material of some sort, and their occasional moisture encourages the growth of mosses, algae and lichens on their steep walls. A few such potholes, however, are so located that the frequent desert winds keep them scoured clean of life.

Potholes, Padre Bay slickrock, Lake Powell.

Such a pothole usually contains a little clean loose sand that acts to sandblast its walls when the wind spirals down into the hole, but the pothole otherwise looks new and recently "drilled". Potholes of small or moderate size can be found where the upper surfaces of harder sandstone strata are exposed, but are especially frequent in deposits of desert-dune and beach-sand origin. Exposures of Navajo Sandstone invariably contain numerous potholes. Some especially large potholes can be seen in the tops of the monolithic sandstone peninsulas that jut south into Padre Bay, on Lake Powell, and in the top of the Navajo Sandstone that walls the Colorado River gorge just downriver from Moab Valley. The Moab Rim and Poison Spider Mesa off-road vehicle trails go within short hiking distances of these eroded rimlands. Other very large potholes in canyon country have even played a part in the pioneer history of the region.

Rincons. "Rincon" is a Spanish word meaning corner or nook. In canyon country, the term is used in two ways. It is applied to any short tributary or alcove in a larger canyon or valley, but in this application is not uniquely descriptive. It is also applied to abandoned loops of entrenched river gorges.

Such loops of an entrenched meander are abandoned when a river finally cuts through the rock wall that separates two curves in the gorge that closely approach each other. Once the cut is complete, the river takes the resulting shortcut and abandons the longer loop. In time, this loop may partly fill with sediments carried in by wind and water, and the river may cut its active bed to depths far below the bottom of the abandoned meander.

The Rincon, a few miles uplake from the Escalante arm of Lake Powell, is an example of such an abandoned meander. A Colorado River rincon more recently abandoned is called Jackson Hole. It is downriver from Moab Valley and can be entered via the Jackson Hole off-road vehicle trail, or viewed from above from the Amasa Back trail. A similar rincon of the Green River can be visited by boat and hiking at the mouth of Horseshoe Canyon, to the northwest of Canyonlands National Park. There are numerous smaller rincons in other canyons throughout canyon country. Many of the larger ones are shown on regional topographic maps.

The Rincon, Colorado River gorge, Lake Powell.

132

Ripple rock, Hellroaring Canyon.

Ripple rock. Ripple rock is rock that was originally sediments rippled by water flow. The ripple pattern was buried under other sediments with the whole geologic layer later buried and turned to rock. When exposed again by erosion, such rock tends to part or delaminate along planes of differing sediment texture, revealing layer after layer of petrified water ripples.

Scientists can determine several things about the original water flow, such as direction, depth, speed, etc., from ripple rock patterns. Ripple rock occurs in several canyon country strata, but is especially abundant in the Moenkopi Formation.

River gravel. River gravel is common enough almost everywhere else, but in a predominantly sandstone region such as canyon country it is more of a curiosity, for several reasons. First, most river gravel is harder rock that has been rounded by abrasion in flowing rivers and streams. In canyon country, most such rock has been carried into the region from elsewhere. Sandstone is generally too soft, and sensitive to decomposition on contact with water, to survive long in smaller pieces. In flowing rivers and streams, such pieces generally disintegrate quickly into sand.

Thus, most canyon country river gravel is from elsewhere, or has been exposed in the few conglomerate rock strata by erosion. Because of these factors, river gravel in canyon country, whether of recent or ancient origin, is often found in strange places, such as lying around on top of exposed sandstone surfaces. Such deposits, if found near a river, may well have been left there by the river before it cut to its present depth. Such abandoned river gravel may also be darkened by desert varnish if it has been exposed to weathering for a long time.

Great deposits of river gravel are also left in abandoned river meanders, or rincons. Jackson Hole, downriver from Moab Valley, contains millions of tons of such "foreign" rock.

Sand dunes. In canyon country, a desert region where most of the exposed rock strata are sandstone, there should be a lot of sand around. There is. There should also be plenty of shifting sand dunes, but there are actually very few. As noted earlier, microscopic plant communities build a protective crust on the sandy soil, thus drastically slowing wind and water erosion. Even so, water erosion does carry away loose sand and other small particles with every rain, especially where human activities have damaged or destroyed the protective cryptogamic soil, and winds do have free access to enough sand particles in some areas to create dunes.

These dunes fall into three general categories. One type forms on the windward and leeward sides of large rock masses or cliff faces, where sand carried by the prevailing winds is deposited in steep

Sand dunes, White Wash.

134

slopes at the bases of such rises. Generally, such dunes grow very slowly, but never shift or move, and vegetation slowly covers them with the aid of a cryptogamic crust. Such stabilized, or partly stabilized, dunes are fairly common in much of canyon country.

A second type of dune occurs in open desert. These dunes are more or less conventional in shape, but their movement is severely limited by low shrubs and grasses that grow among them. There may be a few patches of dunes in such areas that have some limited freedom of movement, but most are effectively restrained by tenacious desert vegetation.

The third type of dune also occurs in open desert, but is quite rare in canyon country. These are the usual rolling, shifting sand dunes that dominate large areas and are virtually unrestrained by vegetation. One such area in canyon country that is especially beautiful is in White Wash, an open desert tributary of the Green River, to the south of the town of Green River. There, decomposing outcrops of salmon-hued Entrada sandstone contribute sand and color to hundred of acres of shifting dunes. These are slowed in their movement only by the rock outcroppings, a few scattered cottonwood trees and sparse patches of desert shrubs. This area is accessible via the Floy Wash Road and White Wash off-road vehicle trail.

Slickrock, Rainbow Rocks.

Slickrock. In canyon country, this term is commonly applied to exposed masses of relatively hard sandstone, in particular the kinds of sandstone that erode into great, rounded monolithic shapes and surfaces, such as the Cedar Mesa, White Rim, de Chelly, Wingate, Navajo and Entrada sandstones. In other parts of the West, exposed and weathered granite-type rock may be called slickrock.

The origin of the term "slickrock" will no doubt be argued endlessly by historians, but it was probably coined by pioneers who found almost any kind of solid rock to be "slick" to their steel-rimmed wagon wheels, and their steel-shod horses' hooves. To modern man, canyon country slickrock is misnamed. Even when wet, sandstone provides good traction for rubber tires and rubber-soled shoes and boots. Only when it is covered with snow does canyon country slickrock live up to its name. Then, traction is virtually nil for any kind of wheel or shoe sole.

Spires. There are three general types of spires in canyon country, each representing different types of rock and their methods of erosion.

One type, represented by the slender spires of Monument Valley, forms from thick strata of desert-dune sandstone, such as the de Chelly, Wingate and Entrada sandstones. Three factors contribute to the spire-forming characteristics of these sandstones: they are all based on softer strata that erode away gradually; they are mechanic-

Fisher Towers.

ally weak and cannot support their own weight when undercut; and when undercut, they scale away in nearly vertical planes. These factors combine with erosion to cause sheer clifflines of these sandstones gradually to recede, and slender walls to become rows of spires before collapsing entirely.

Navajo Sandstone, although it has two of the three spire-forming characteristics, lacks the third, a soft base. The harder, horizontal layering of the Kayenta Formation beneath the Navajo does not erode away gradually. Rather, as its softer layers erode away, the harder layers settle in large, irregular slabs. This difference in how its "foundation" fails is the key to why Navajo Sandstone rarely forms slender spires. It is also the key to why magnificent Rainbow Bridge, a span formed entirely of Navajo Sandstone, will someday collapse suddenly, catastrophically, without warning, its foundation of Kayenta weakened by wetting from standing Lake Powell water.

The second type of canyon country spires forms from such formations as the Cutler, Moenkopi, and, to some extent, the Honaker Trail, Carmel, and Dewey Bridge, plus local areas of several other strata. This type of rock tends to be eroded by rain into vertical towers with surfaces convoluted because of layers of differing hardness. Generally, caprocks, or harder uppermost slabs of rock, help rain erosion shape the slender spires.

The Fisher Towers, near Utah 128 to the northeast of Moab Valley, are outstanding examples of this type of spire. Others occur almost everywhere the noted formations are exposed to any extent. A few can be seen to the west of U.S. 163 about five miles north of the Colorado River. There are others in Arches National Park north of the Courthouse Towers area. Various off-road vehicle trails in eastern canyon country, such as the Onion Creek, Hurrah Pass, Lockhart Basin and Potash trails, go through or near groups of these colorful, oddly shaped spires.

The third type of spire is rare in occurrence because it depends upon a combination of fairly erosion-resistant rock and a pattern of fracturing that is not common. One outstanding area of such spires is the "needles" in Canyonlands National Park. Here, the Cedar Mesa Sandstone has been jointed and faulted in a close, cross-hatched pattern by the Monument Uplift and by subsequent slumping during graben forming. The resulting slender, vertical slivers of rock have eroded along their fracture lines to form rows and groups of sharp-tipped, color-banded "needles," or spires. The Elephant Hill off-road vehicle trail provides excellent views of these spectacular spires.

Springs. Canyon country springs that occur in mountainous areas are much the same as mountain springs anywhere, but the numerous springs that appear in the lower sandstone country are oddities. As noted earlier under "canyon streams," such desert springs appear when subsurface water, seeping through porous strata, are channeled to the surface by layers of impermeable rock. When such springs

Spring, Cane Creek Canyon.

appear in canyon walls, they often form spring-seep caves or alcoves. When they do not encounter steep slopes or cliffs, they may surface as seeps in relatively level terrain, there to trickle off into the desert to evaporate, or disappear into the first stretch of permeable ground.

Canyon country desert springs, whether in canyons or cliffs, on steep slopes or in open desert, are always marked by patches of water-loving vegetation, especially the exotic salt cedar, or tamarisk shrub, with its willowy fronds, and tiny flowerlets.

Tar sands. Tar sand is sandstone that is saturated with petroleum from which the lighter, more volatile compounds have escaped, or evaporated. There are enormous amounts of tar sand in northeastern Utah, but canyon country also has areas where this strange tar-like material is found.

The largest such deposit is in Elaterite Basin, just west of The Maze, in Canyonlands National Park. There, what was once an enormous off-shore sand bar in White Rim Sandstone, is now exposed by erosion. Before exposure, that elongated, underground dome of permeable sandstone was filled to saturation with oil that was being forced upward by water. Since the sandstone dome was roofed by impermeable shales, the oil was trapped.

As the oil-filled dome was gradually exposed by erosion, the more volatile components of the oil were cooked out by the desert heat, leaving a heavy tar-like oil that even today seeps from the rock many places on hot days. An off-road vehicle trail in the vicinity of the Maze District of Canyonlands National Park traverses Elaterite Basin and provides access to its tar sand outcropping.

There are other smaller tar sand deposits within the San Rafael Swell.

Tunnels. Tunnels are one of the four basic types of natural openings in rocks that occur in canyon country. The other three are arches, bridges and windows. Essentially, a tunnel is an elongated hole through a rock mass, generally through a thick, monolithic type of rock such as Entrada, Navajo, Wingate or Cedar Mesa sandstone. How short the tunnel can be before it becomes an arch or window is a matter of detailed definition, and will vary from expert to expert.

There is an excellent example of a tunnel-type opening in the Devils Garden vicinity of Arches National Park. Whoever named this opening evaded the problem of precise definition by calling it Tunnel Arch. This kind of tunnel forms when a pothole drills entirely through a mass of rock at a shallow angle, possible following a compositional anomaly in the rock.

Tunnel-type openings that are clean holes through solid rock are rare, but another type that forms when two spring-seep caves meet is more common. Examples of this type can be found near Pritchett Arch and near The Bride spire, in a tributary to Little Canyon, near the Gemini Bridges off-road vehicle trail.

Manmade tunnels in the form of abandoned mine shafts are also fairly common in canyon country, but are not known for their aesthetic beauty. They are also very hazardous to enter.

Uranium. Uranium in the form of several different chemical compounds occurs in various canyon country rock strata. Geologists believe that this radioactive element was leached by water from the igneous rock thrust upward through the region's sedimentary strata when its laccolithic mountains formed. Soluble uranium salts have the curious property of changing and precipitating out of water solution as insoluble oxides and other compounds when they come in contact with organic materials. Thus, uranium concentrations are often found in rock strata that contained a lot of ancient plant debris. Petrified logs heavy with yellow uranium compounds are sometimes found, and petrified dinosaur bones containing uranium also occur in certain strata.

Uranium compounds occur in traces in several strata, but are especially likely in the Morrison and Chinle formations. These freshwater floodplain deposits were heavily vegetated.

Uranium mining in canyon country has enriched its human economy, but has blighted its unparalleled beauty in many places with prospecting and mining activities. Such activities first boomed in the region during the 1940s and '50s, but have recently renewed manyfold and continue today.

It should be noted that uranium ores are radioactive and often contain chemical compounds toxic to humans. Abandoned uranium mines are especially hazardous, from both safety and health viewpoints. Only those who are prepared to endanger their long-range health should expose themselves to uranium, its ores, sources and milling wastes, all of which are found in canyon country.

Windows. Windows are one of the four basic types of natural rock openings. As noted earlier, a window can roughly be defined as a hole through a relatively thin wall of rock, with the hole so far from any edge of the wall that it does not give obvious shape to the rock between the hole and edge. If the hole should lend shape to an edge, such as a recognizable arc, then the opening might be called an "arch" rather than a window. Or if the rock wall is thick enough, the opening might be classified as a "tunnel." The difficulty of objectively resolving such fine points of definition is one thing that prevents standardization of rock-opening nomenclature.

Window openings form in the same manner as arch openings. They start in thin rock walls, either by water erosion at a point where cementing minerals are more soluble, or where internal rock stress has forced rock to crack away. Surface erosion processes then slowly round and enlarge the opening.

There are many window-type openings in Arches National Park. A small one can be seen from Arches Road in the Courthouse Towers

Window Arch, Pritchett Canyon.

area. In other areas, names and definitions do not always correspond. For example, along the trail to Double-O Arch, Black Arch, Partition Arch and others might more properly be termed "windows," and in the Windows Section of the park, both the North Window and South Window meet most definitions of arches.

Most of the windows in Arches National Park are in Entrada Sandstone and were probably started by water erosion. Window Arch, in upper Pritchett Canyon, is a good example of a window that probably started by collapse caused by internal rock stress. The Pritchett Canyon/Behind-the-Rocks off-road vehicle trail goes by Window Arch.

Wrinkled rock. The phenomenon of "wrinkled rock" is not common in canyon country, but is quite distinct in the general vicinity of Moab and at scattered sites in the back country south and west of Bluff. In the Moab area, wrinkled rock occurs in the Dewey Bridge member of Entrada Sandstone, the dark red layer of mudstone that lies between Navajo Sandstone and the Slickrock member of Entrada Sandstone. In the southeastern corner of canyon country, the phenomenon affected the equivalent Carmel Formation.

The wrinkling that affected both the Dewey Bridge and Carmel mudstones caused their horizontal layering to be severely distorted into undulating waves that are clearly visible where the edges of these strata are revealed by erosion. In most locations, some of the overlying Entrada Slickrock sandstone was also affected by the wrinkling process, but the underlying Navajo Sandstone was not.

Geologists speculate that this mysterious wrinkling occurred when the level mudstones had been covered to some depth by Entrada desert sands. Then, for some as yet undetermined reason, the affected areas were tilted slightly. This allowed the still wet, pliable, uncemented mud strata, with their overburden of sand, to slip slightly downhill, sliding on the eroded upper surface of the better cemented Navajo Sandstone, and producing the wrinkled effect now visible in the exposed rock.

This wrinkled rock can be viewed in many places, but two sites are easily accessible. In Arches National Park, the phenomenon can be seen from the paved Arches Road beginning at the park entrance station, and at intervals on to the road's end in the Devils Garden area. It is striking in the Windows section that is entered by a spur road. Another outstanding exposure can be seen from Utah 313 after it has climbed out of Sevenmile Canyon. The conspicuous Monitor and Merrimac buttes to the north of the highway, and the cliffs on west, all have wrinkled Dewey Bridge mudstone at their bases, and white terraces of Navajo Sandstone below that.

ROCKHOUNDING

GENERAL

Early gold prospectors had a saying, "gold is where you find it," meaning that its location is difficult to predict on the basis of visible geologic clues. The same phrase and difficulty could also be applied to the collection of mineral specimens in most regions, but not in canyon country.

In canyon country, it is possible to apply a general knowledge of the region's geology and produce the desired specimens almost every time. Further, canyon country often rewards rockhounding diligence with exciting serendipitous discoveries, such as petrified prehistoric animal tracks or bones, or petrified botanical specimens or impressions.

Rockhounding, Poison Spider Mesa.

In most desert regions, millions of years of erosion have turned rockhounding into a hit or miss activity, with worthwhile specimens widely scattered or limited to small outcrops. Canyon country has also been eroded, so violently in the last 10 million years that it has lost thousands of cubic miles of rock. This rapid cutting, however, has turned the remaining desert-canyon country into an enormous outdoor "supermarket" for rockhounds, with all the collectible specimens laid out neatly on "shelves."

As in any supermarket, however, it is necessary to know where to look for what is desired. It is therefore necessary for the serious canyon country rockhound to learn how to identify the various "shelves," or geologic strata, and what kinds of specimens occur in each.

An earlier chapter discussed the various rock strata in canyon country and provided some information useful in identifying them in the field. Geologic maps that are color coded to show what rock formations are exposed on the surface aid in strata identification and are the key to systematic, successful mineral specimen collection. Some sources for such maps are listed under Further Reading.

Detailed information about what types of minerals occur in various canyon country rock strata can be found in other publications, but a partial list is provided in the following section of this chapter. As noted under "river gravel" in the previous chapter, rocks from outside of the region are also found, where abandoned by rivers. Such sites offer rockhounds a wide variety of specimens from which to choose, all within a small area.

In practice, rockhounds can study the following list, as augmented from other sources, and find which rock strata contain the types of specimens they are seeking. The next step is to study a geologic map that shows what formations are exposed on the surface, and locate accessible exposures of the formation selected. Canyon country road and trail maps provide access information in the general Moab vicinity, and various other maps provide limited information about other areas. Some sample exposures of the various rock formations are listed in the earlier chapter on rock strata, and some representative collecting sites are listed in the last section of this chapter. Other specific collecting sites are listed in books noted under Further Reading.

Those who visit canyon country for the purpose of collecting mineral specimens should keep the following limits in mind:
1.   It is not legal to collect specimens of any sort within national or state parks or monuments. Limited collecting is permitted within national recreation areas. For current regulations in recreation areas, check with park rangers.
2.   Both federal and state laws prohibit collecting or disturbing "antiquities." This includes archeological and historical remnants and the recognizable remains or imprints of prehistoric animals. The non-commercial collecting of small amounts of petrified wood and marine specimens is generally permitted on most non-park public land, but

144

Rockhounding, Klondike Bluffs.

current regulations prohibit collecting intact petrified bones of prehistoric animals.

3. Discoveries of prehistoric Indian remnants, or petrified animal bones or plants, should be reported to the appropriate land administration agency. For most of the region, this is the Bureau of Land Management, which has offices in most communities.

It should be noted that not all the canyon country geologic strata listed in the earlier chapter on rock strata contain collectible mineral specimens, and others are either not exposed on the surface or are not commonly accessible. Rock strata not listed in the following section fall into one of these three categories. Those strata which do contain collectible minerals or interesting fossil traces are listed, from oldest to youngest.

MINERALS IN VARIOUS STRATA

Paradox Formation. Primarily shallow marine salts, but with dolomite, gypsum crystals and a few marine fossils in some exposures.

Honaker Trail Formation. Abundant marine fossils and some petrified wood.

Moenkopi Formation. Abundant ripple rock and mud casts, with a few marine fossils in the western areas of canyon country, and some tracks and fossilized reptile and amphibian bones in the eastern areas.

Chinle Formation. Agate, petrified wood, calcite and barite geodes, jasper, aragonite, selenite, fossil ferns and horsetail, with tracks, petrified bone, coal, jet, carbonized plantlife and marine fossils in some localities.

Kayenta Formation. Ripple rock, mud casts, iron concretions and dinosaur tracks.

Navajo Sandstone. Chert, iron concretions, rare petrified bone and the tracks of small to medium animals in desert playa layers.

Rockhounding, beyond Hurrah Pass.

Carmel Formation and Dewey Bridge member, Entrada. Agate, chert, iron concretions and some petrified wood.

Curtis Formation. Marine fossils.

Summerville Formation. Agate, chert, jasper, gypsum, ripple rock, mud casts and animal tracks.

Morrison Formation. Agate, chert, petrified wood and bone, plant and invertebrate fossils, calcite, barite, celestite, selenite, geodes, gastroliths, coprolite and dinosaur tracks.

Cedar Mountain and Burro Canyon formations. Some plant and invertebrate fossils, with rare petrified bone and wood, fossils of fresh water mollusks, quartz crystals and fish scales.

Dakota Formation. Agate, chert, fossilized bone, plants and coal, with abundant petrified wood in some localities.

Mancos Shale. Calcite crystals, aragonite, marine fossils, coal, jet and carbonized plant life, with some fossil plant impressions in upper layers.

Igneous exposures. Exposures of igneous rock in the Abajo, Henry and La Sal mountains contain various copper minerals, silver and gold, with some smokey quartz in the La Sals. Some hornblend and placer gold occurs in the Henrys.

SAMPLE LOCATIONS

The following sample sites for mineral specimen collecting are merely examples. There are many other major sites and countless smaller exposures of the geologic strata that contain mineral and fossil specimens of interest to collectors.

Access instructions to the various sites listed are based upon backcountry road and trail nomenclature standardized in other Canyon Country guidebooks, as listed in the back of this book. All of the sites listed are on non-park public land.

1. *Little Valley.* Drive north from Moab on U.S. 163 to the Klondike Bluffs trail, then east on that trail into Little Valley. The western slopes and hills that border this nine mile long valley are the Morrison and Cedar Mountain formations.

2. *Little Canyon.* Drive north from Moab on U.S. 163 to the Gemini Bridges trail, then west on that trail into Little Canyon. There are good exposures of the Chinle Formation in the eastern end of this "hanging canyon."

3. *Monitor and Merrimac.* Drive north from Moab on U.S. 163 to the Monitor and Merrimac trail, then west and south on that trail to the Monitor and Merrimac buttes and nearby bluffs and spires. There are unusual chert and agate specimens in the Dewey Bridge member at the bases of these Entrada Sandstone monoliths.

4. *Pyramid Butte.* Drive north from Moab on U.S. 163, then west on Utah 279 to its end. Continue on the Potash trail to beyond the potash ponds, then turn south on the Pyramid Butte spur trail. The upper Honaker Trail Formation is exposed along this spur trail to its end.

5. *Yellowcat.* Drive south from Interstate 70 on Yellowcat Road, then explore various trails in the Yellowcat area, where there are exposures of Mancos, Dakota, Morrison and Summerville strata.

6. *San Rafael Valley.* Drive south from Interstate 70 on Utah 24. From the junction south for several miles there are exposures of Morrison sediments on both sides of the highway.

7. *Salt Wash.* Drive southwest from Interstate 70 on Floy Wash Road, then north on the Salt Wash trail. This trail travels through Summerville sediments, with Morrison on the higher ground on each side of the wash.

8. *Spring Canyon Point.* Drive north out of Moab on U.S. 163, west on Utah 313, north on Dubinky Well Road then west on the Spring

Canyon Point trail to its end on Spring Canyon Point. There, large deposits of ancient river gravel offer a variety of minerals.

9. **San Rafael Swell.** There are many excellent rockhounding areas in the western and northern areas of the San Rafael Swell. Various U.S. Geological Survey topographic maps and Utah state recreation maps show backcountry access roads and trails within this uplift.

10. **Henry Mountains.** The foothills areas of the Henry Mountains offer many excellent rockhounding sites. U.S.G.S., Bureau of Land Management and Utah state recreation maps show the access roads and trails in this area.

11. **Lake Powell.** Navajo Sandstone dominates much of the shoreline of this reservoir. In many places, remnants of the Carmel Formation and ancient river gravel deposits can be found just above the Navajo, or isolated on the tops of Navajo domes. Such deposits offer varied rockhounding opportunities. Stream-rounded pieces of petrified wood are sometimes found in the gravel deposits.

Marine fossils, Colorado River gorge.

INDEX OF
SURFACE FEATURES

Entrada Sandstone fins, Arches National Park.

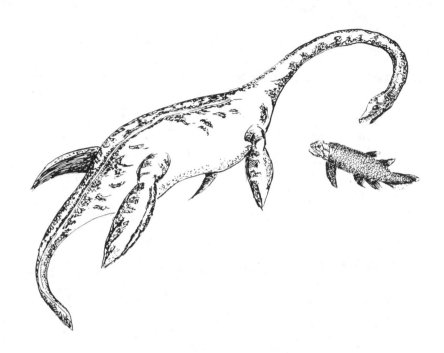

FURTHER READING

GENERAL

As an aid to the selection of books and maps for further reading, those that are intended primarily for serious students of geology are coded (T) for technical. Books that are intended primarily for non-technical readers are coded (P) for popular. Those useful to both are coded (P/T).

GEOLOGY TEXTBOOKS

Principles of Geology, Gilluly, Waters and Woodford, W. H. Freeman and Company. (T)

The Evolving Earth, Sawkins, Chase, Darby and Rapp, Macmillan Publishing Company. (T)

Physical Geology, Flint and Skinner, Wiley and Sons, Inc. (T).

Geologic Time, Eicher, Prentice-Hall, Inc. (T)

Dictionary of Geological Terms, American Geological Institute, Anchor Press, Doubleday. (P/T)

CONTINENTAL DRIFT

A Revolution in the Earth Sciences, Hallam, Clarendon Press, Oxford. (T)

Continental Drift, Tarling, Anchor Press, Doubleday. (P/T)

The Restless Earth, Calder, Viking Press. (P/T)

Continental Drift and Plate Tectonics, Glen, Merrill. (P/T)

Mesozoic and Cenezoic Paleocontinental Maps, Smith and Briden, Cambridge University Press. (P/T)

Drifting Continents, Shifting Seas, Young, Franklin Watts. (P)

REGIONAL GEOLOGY

The Evolution of North America, King, Princeton University Press. (T)

Utah, Muench and Wixom, Belding. (P)

Slickrock, Abbey and Hyde, Sierra Club. (P)

Standing Up Country, Crampton, Knopf. (P)

Red Rock Country, Baars, Natural History Press, Doubleday. (P/T)

Geologic History of Utah, Hintze, Brigham Young University, Department of Geology. (T)

Guidebook to the Four Corners, Colorado Plateau and Central Rocky Mountain Region, National Association of Geology Teachers, Field Conference, 1970. (T)

Guidebook to Canyonlands Country, Four Corners Geological Society, Eighth Field Conference, 1975. (T)

The Geologic Story of Canyonlands National Park, Lohman, U.S. Geological Survey Bulletin 1327. (P)

The Geologic Story of Arches National Park, Lohman, U.S. Geological Survey Bulletin 1393. (P)

The Geologic Story of Colorado National Monument, Lohman, U.S. Geological Survey, Black Canyon Natural History Association. (P)

Scenes of the Plateau Lands, Stokes, Publishers Press. (P)

Geology of Canyonlands and Cataract Canyon, Baars and Molenaar, Four Corners Geological Society, Sixth Field Conference, 1971. (P)

Geology to the Canyons of the San Juan River, Baars, Four Corners Geological Society, Seventh Field Conference, 1973. (P)

Guidebook to the Colorado River, Part 3, Moab to Hite, Rigby, Hamblin, Matheny and Welsh, Brigham Young University, Department of Geology. (P)

River Runners Guide—Desolation and Gray Canyons, Mutschler, Powell Society, Ltd. (P)

River Runners' Guide—Labyrinth, Stillwater and Cataract Canyons, Mutschler, Powell Soceity, Ltd. (P)

River Runners Guide—Canyonlands National Park and Vicinity, Mutschler, Powell Society, Ltd. (P)

MAPS

"Geologic Map of Utah, Southeast Quarter," Hintze and Stokes, Utah Geological and Mineralogical Survey. (P/T)

Utah Geological Highway Map, Hintze, Brigham Young University, Department of Geology. (P/T)

Geological Highway Map, Utah, Colorado, Arizona and New Mexico, American Association of Petroleum Geologists. (P/T)

Geology, Structure and Uranium Deposits of the Moab Quadrangle, U.S. Geological Survey. Other areas of canyon country are covered by this series on quadrangles entitled Grand Junction, Cortez, Shiprock, Marble Canyon, Escalante, Salina and Price, although some of these may now be out of print. (P/T)

Utah Multipurpose Map #1, Southeastern Utah, Utah Travel Council. (P)

Utah Multipurpose Map #2, Southeastern Central Utah, Utah Travel Council. (P)

ROCKHOUNDING

A Field Guide to Rocks and Minerals, Pough, Houghton Mifflin Company, Peterson Field Guide Series. (P/T)

A Collector's Guide to Mineral and Fossil Localities in Utah, Utah Geological and Mineralogical Survey. (P)

Minerals of Utah, Bullock, Utah Geological and Mineralogical Survey. (T)

Mineral Tables, Dietrich, McGraw-Hill. (P/T)

Utah Gem Trails, Simpson, Gem Trail Publications. (P)

Fossils for Amateurs, MacFall and Wollin, Van Nostrand Reinhold. (P)

MISCELLANEOUS

Field Guide to Landforms in the United States, Shimer, Macmillan Publishing Co. (P/T)

1,000 Million Years on the Colorado Plateau, Look, Bell Publications. (P)

Geologic Time, Newman, U.S. Geological Survey. (P)

Colorado River gorge.

Notes